CRISIS

MANAGEMENT

IN CONSTRUCTION

PROJECTS

MARTIN LOOSEMORE, PH.D.

American Society of Civil Engineers
1801 Alexander Bell Drive
Reston, Virginia 20191–4400

Abstract: This book focuses on preventing crises and, if that's not possible, turning them to advantage. It draws lessons from a range of industries, concluding that the secret of effective crisis management is balancing prevention with control. Unfortunately, crises have built-in defense mechanisms that cause people to act in ways that makes this difficult. The book also gives advice as to how to overcome the built-in defense mechanics that cause people to act in way that may impede effective crisis management.

Library of Congress Cataloging-in-Publication Data

Loosemore, Martin.
  Crisis management in construction projects / Martin Loosemore.
     p. cm.
Includes bibliographical references and index.
  ISBN 0–7844–0491–7
  1. Building—Safety measures. 2. Emergency management. I. Title.

  TH443 .L75 2000
  690'.068'4—dc21

                                        00-064569

To Tristan, Bryony, and Elliot

TABLE OF CONTENTS

ACKNOWLEDGMENTS

Thank you, Heather, for your unending patience and support. Thanks also to my three children, without whom this book would have been much easier but far less worthwhile!

There are many other people I need to thank such as my parents, John and Anita, who have taught me so much and, the colleagues who have helped shape my ideas. In particular, I would like to thank Peter Hibberd, Will Hughes, Trevor Francis, Denny McGeorge, Goran Runeson, Melissa Teo and Derek Walker for their friendship, criticism, and support.

Finally, thank you to my colleagues at The University of New South Wales, for providing a supportive environment in which to write this book. It is a pleasure to work in a faculty that is so conducive to research.

PREFACE

Political, economic, and social instability; depleted natural resources; increasing global competition; and rapid technological advances are making business increasingly unpredictable. Unexpected problems are the norm rather than the exception and high-profile engineering disasters such as the Hyatt Regency Hotel Walkway collapse in Kansas City in 1981 are merely the tip of the iceberg. Risk theory suggests that for every reported crisis, the multipliers in terms of unreported incidents are enormous. For example, Smith (1996) indicates that for every fatality in the airline industry there are 10 major accidents, 30 minor accidents, and 600 near misses. This book draws an important distinction between the day-to-day problems that constantly punctuate the lives of managers and the occasional crises. The challenges of managing crises demand special attention because they hyper-extend organizational systems and personnel, posing managers with an extraordinary array of complex problems that can rapidly escalate into full-blown disasters. This book is about preventing this from happening.

One of the unique aspects of this book is its *preventative* and *reactive* focus. This contrasts with traditional construction management texts that have been dominated by prevention strategies rather than strategies for dealing with crises when they occur. While prevention is better than cure, it is increasingly unlikely that managers can create a crisis-free environment. This requires that organizations have reactive capabilities to deal with the unexpected.

Another unique feature of this book is its focus on *people.* This, too, is in contrast to traditional construction management texts, which are essentially scientific in their approach, being characterized by a plethora of bar charts, networks, and cash-flow graphs. While these texts emphasize measurement, control, and universal prescription as a means of reducing uncertainty, this book emphasizes thoughtfulness, flexibility, and the accommodation of uncertainty. This alternative approach to construction project management is important because there has been precious little research into the human aspects of construction project management. As Butterfield (1975) argues, "no field of thought can be properly laid out by men who are merely measuring with a ruler" (page 1).

Who is this book written for?

Crises have no respect for professional boundaries, which makes this book relevant to *all* who manage construction projects. While it has a construction flavor, it draws lessons from a wide range of industries and therefore, should also be of wider interest.

Although this book is primarily designed for practitioners, it challenges the traditional assumption that theory and practice are incompatible. Watson (1994) has shown how practicing managers intuitively develop, update, and draw on complex theoretical ideas in their day-to-day lives. Although they may not be formal management theories, managers are essentially "practical-theorists" and the most effective of them recognize the value of theories as conceptual frameworks to guide their actions. For this reason, this book marries an applied orientation with elements of strategic and basic research.

Distinctive text features

Experience is important when dealing with crises, but getting that experience is difficult and painful. The best substitute is to relay the experiences of others through case studies of past crisis management efforts. But, there are very few detailed accounts of crisis management in a construction context. To redress this deficiency, this book analyzes four actual construction crises. These case studies provide an excellent vehicle for readers to relive other peoples' experiences and the insights they provide are almost boundless. In particular, they help to develop an understanding of *how* and *why* people behaved as they did during a real-life crisis, thereby enabling judgments to be made about the influence of certain behaviors on crisis management outcomes.

In an educational sense, the case studies also provide interesting and amusing teaching material for analysis and discussion. The managerial issues covered include problem solving, decision making, leadership, communication, risk management, disaster management, health and safety, conflict and change management, organizational design, and teamwork.

Martin Loosemore, Ph.D., M.ASCE, MCIOB
Associate Professor
University of New South Wales
Sydney, Australia

Chapter 1

Introduction

This chapter challenges the anti-conflict value-system that causes most people to see crises as threats rather than opportunities. It demonstrates how well managed crises can strengthen projects rather than destroy them.

OUR ANTI-CONFLICT VALUES

To dedicate an entire book to crisis management may seem negative and defeatist. This is understandable in the context of modern business trends such as total quality management, partnering, and business process re-engineering. Such trends are increasingly evident within the construction industry and teach managers that prevention is better than cure, that crises are a sign of managerial failure and that any form of disharmony or conflict is wasteful and damaging.

The origin of these anti-conflict values lies in the institutions that indoctrinate us during our most impressionable and formative years. For example, at home we are often taught that our parents "know best," that we should "not answer back," and that we "should be seen and not heard." Church doctrines tell us that happiness is found in togetherness, harmony, peace, and tranquillity. At school, teachers have the right answers and are not to be challenged. Furthermore, traditional approaches to teaching tend to suppress any creativity and critical inquiry that children might possess, teaching them that the secret of success is giving the answers that coincide with what the teacher believes to be true. Indeed, as we mature, the institutional pressures to conform do not subside. For example, entry into a profession normally requires passing demanding examinations and subscribing to strict codes of professional practice. This system is designed to maintain and perpetuate the traditional roles that each profession has carved out for itself, which serve to distinguish its members from other professions and to define their status in society.

Many of the most vivid examples of institutions constraining the way we think can be found throughout the development of modern science where the church in particular, held back advances in understanding. For example, in the ecclesiastical universities of the middle ages, the belief was that it was necessary to separate the pursuit of truth from mankind's day-to-day cares and that the church was the high protecting power of all intellect, discovery, knowledge, science and speculation. While it is a fallacy that the church regularly punished and tortured those who criticized its doctrines, scientists, most of whom were monks, were expected to refrain from directly and explicitly challenging its authority. For example, although letters show that the church encouraged Copernicus to publish his theory that the sun rather than the earth was at the center of the universe, there is also evidence that he delayed its publication for 30 years because he was tortured by the inevitable criticism it would bring from his peers (Koestler 1975). Indeed, in 1633, when Galileo published research that supported Copernicus' earlier theories, he was forced to retract his findings before a Catholic church tribunal. Incredibly, the charges against him were only recently annulled by the Pope.

While we have come a long way since the middle ages, modern institutions still govern our lives and it is not surprising that two of mankind's most profound advances in knowledge, the theory of relativity and the theory of evolution, occurred largely outside their influence. It is interesting to note that both Einstein and Darwin were amateurs who were not strongly affiliated to any particular university or church. This strong sense of individualism is also recognizable in many of the most influential political thinkers of our time. For example, Margaret Thatcher is noted for saying that "there is no such thing as society" and Aleksandr Helzen, who developed the concept of Glasnost which was later adopted by Mikhail Gorbachev, lived in exile for most of his life and rejected any form of collective action that might have compromised his beliefs. It would seem that disharmony is not as undesirable as we are taught and that creativity often comes from those who challenge the powerful anti-conflict values which pervade society and shape our behaviour.

NECESSITY—THE MOTHER OF INVENTION

The anti-conflict values that are increasingly propounded in mainstream management and in construction management are worrying since they create organizations that are driven by strategies of mitigation rather than optimization. Such organizations are incapable of achieving their full potential. While this book may initially appear negative, its aim is to challenge this pessimistic mind-set by developing a framework for an optimistic organization. In contrast to pessimistic organizations, optimistic organizations are driven by values that see crises as potential opportunities rather than threats, which, if well-managed, can enhance rather than threaten their effectiveness.

The case for optimism is perhaps most vividly illustrated in our history books, which are littered with examples of people demonstrating their greatest capacity for ingenuity in the face, midst, or aftermath of crises. Epidemics, in particular, have provided fertile ground for innovation. For example, the Black Death prompted quarantine regulations, Small Pox led to vaccinations, and Cholera was the impetus for the first public health authorities. Major disasters have also stimulated our creative capacities. For example, the great fire of London in 1666 gave architects like Christopher Wren and Nicholas Hawksmoor the chance to build many beautiful cathedrals such as St Paul's and also provided Britain with its first building regulations. Similarly, the Great Fire of Chicago in 1870 provided architects and engineers with the chance to construct the world's first steel-frame skyscrapers. Indeed, Chicago is still renowned for its innovative and exciting architecture.

In modern times, the most prominent stimulant to invention has been two world wars, the second of which led to the development of radar, DDT, Penicillin, jet-propelled aircraft, the United Nations, the World Bank, and the International Monetary Fund. More recently, the cold war gave us nuclear power and the space race, out of which came the first communications satellites and manned space craft.

The field of management has also generated many innovations in the face of adversity. For example, the deprivation, exploitation, and social unrest during the early industrial revolution prompted pioneers like Charles Babbage and Robert Owen to experiment with more paternalistic and socially responsible management practices (Sheldrake 1996). This spurned new ideas such as participatory management, work-groups, the first profit-sharing schemes, and educational and welfare facilities for employees. More recently, the origins of many contemporary management trends that we encounter in construction texts can be traced back to problematic times. For example, value engineering was developed in World War II to cope with severe material shortages that forced manufacturers to identify unnecessary costs in production and to experiment with alternative raw materials. To people's surprise, the results were often cheaper and better quality products (Miles 1967). Total quality management (TQM) can claim the same origin. The momentum for TQM grew out of the economic devastation of Japan during World War II. According to Tsurumi (1982) the destruction imposed on Japan was responsible for the enthusiasm with which the principles of TQM were embraced, coupled with a certain degree of luck in terms of their suitability to Japan's traditional cultural values. As Demming has pointed out in reference to America's continued failure to grasp TQM: "you have to be in a crisis before you pay attention."

Coincidentally, it was Japan's increasing competitive advantage over the Americans during the 1970s and 1980s that gave birth to the modern concept of benchmarking by the Rank Xerox Corporation. When Rank Xerox was challenged by Japanese competitors who could sell copiers for less than U.S. manufacturing costs, they dismantled the Japanese products and used the components as models for their own standards of production (Camp 1989). Later, General Motors and Ford did the same with Japanese gear boxes which were far smoother than theirs, discovering that the tolerances that their competitors were working toward were far smaller than those set by American industry. These principles have now been extended to all areas of business activity in most of the world's leading organizations.

UNLEARNING THE PAST—THE FUTURE'S CHALLENGE

If crises provide the arena for mankind's greatest steps forward, then the future for managers is exciting. In 1970, Alvin Toffler's *Future Shock* portrayed an increasingly chaotic, volatile, and crisis-prone business environment, largely brought about by demographic changes, resource depletion, globalization, technological advances, and world-wide political and economic reform. Indeed, his predictions proved far too conservative and while our predecessors dealt with tried-and-tested technologies in an evolutionary fashion, modern construction managers are faced with an ever-faster flow of new and largely untested technologies, with less time to understand their performance and compatibilities. This increased technical complexity has led to increased organizational complexity, reflected in more specialists with different needs and elaborate interdependencies which are often difficult to understand. To compound these problems, an increasing awareness that the construction industry is a major threat to the health and welfare of its employees, to the general public and to the environment, is leading to increased levels of external regulation. While one can cite past projects as evidence that such pressures have always been present, closer inspection reveals that construction managers have never had to grapple with such a wide variety of pressures, needs, and regulatory constraints. For example, while, in 1931, the program for the 102-story Empire State Building was only 14 months, the technology was relatively well tried and tested, construction regulations were far more lax than they are today, and safety was of secondary concern (Theodore 1975).

Thus, it would seem that the managerial challenge of the future is one of complexity and change rather than shear scale, as it was in the past. As Kanter (1983) noted, there is more competition, more activities to manage and importantly, more limited resources to achieve and fail with. There are also new opportunities, but their exploitation will demand a change toward more thoughtful, flexible, responsive, and human-centered managerial styles. This is because, in an increasingly uncertain and competitive environment, rigid systems become restrictive and counter-productive and an organization's ability to harness the creative capacities of its human resources becomes the basis of its vitality and success (Pascale 1991). The challenge of the future is to use people more effectively to discover better ways of exploiting the exciting opportunities that will arise.

Unfortunately, meeting this challenge has proved difficult in the construction industry because, like many other industries, management continues to be underpinned by the scientific values Frederick Taylor advocated in the early 20th Century (Blockley 1996). As Bea (1994) found in his investigation of marine structure failures, this has meant that most engineers are still very uncomfortable with two things: uncertainty and people. It is a disability that has become an increasing concern within engineering recently and a growing number of engineers are recognizing that their profession's continued prosperity will depend on expanding its focus from "one that deals solely with objects to one that deals equally proficiently with people" (Johns 1999). It is argued that this will better equip engineers to maximize the potential value of an increasingly diverse workforce, ultimately improving the reliability of their buildings and structures.

RECOGNIZING A CRISIS WHEN YOU SEE ONE

One certainty in a world increasingly saturated with complexity and uncertainty is that the future will be punctuated by sudden, unexpected, and potent events that will require a rapid response. These relatively intense events are called crises and while they are less frequent than the day-to-day problems that constantly punctuate the lives of managers, most can expect to have to deal with at least one during their career. The way in which managers deal with such events can mean the difference between corporate life or death and in a construction project setting, minds automatically turn to vivid incidents such as fires, bankruptcy, serious disputes, serious accidents, collapses, strikes, and natural phenomenon such as floods. However, the variety of crises that can arise on construction projects is enormous and in a recent American survey, engineers ranked them in descending order of frequency as: construction delays, design errors; cost over-runs, management successions; local opposition to project; employee raided by competitor; third-party lawsuit; disgruntled employee; merger or acquisition and accidents.

Whatever its precise nature, a crisis is generally accepted to be a low probability, unexpected, high-impact event that is not covered by contingency plans (Booth 1993). Crises represent an immediate and serious threat to high priority goals, placing managers under extreme time pressure to find a non-routine solution. They also have widespread financial, political, cultural and social implications, the consequences of which are likely to be subject to extensive public, media, and/or government scrutiny (Pearson and Clair 1998). Not surprisingly, these extreme characteristics tend to highlight unknown strengths and divisions within organizations, hyper-extending relationships, testing people to the limit, and producing stress and anxiety among organizational stakeholders. In this sense, crises pose managers with very different challenges to their routine day-to-day problems.

DISTINGUISHING BETWEEN DISASTERS AND CRISES

As we have established, crises are complex and unforgiving phenomena that drag high-level decisionmakers into uncharted waters, revealing weaknesses and strengths that would otherwise not be apparent. Most organizations struggle with the complexity and magnitude of the challenges posed by such events, and in contrast to day-to-day problems, the consequences of mismanagement can be "disastrous." This distinction between a disaster and a crisis is important and can cause confusion because the terms are often used synonymously. Essentially, a disaster is the consequence of a mishandled crisis and this book is about managing crises to prevent disasters from happening.

A CRISIS-PRONE CONSTRUCTION PROCESS

The potential for crises during the life of a construction project is enormous. This is due in part, to the increasing technical and organizational complexity that is affecting all industries, but also to the construction industry's peculiar culture and managerial practices.

Cultural problems

The professions

A good point at which to begin a discussion of the construction industry's most significant problems is by referring to the institutionalized divisions that have developed between the occupational groups that contribute to it. To fully understand the nature and significance of these divisions, one must go back to the 19th century, when the process of industrialization fostered the development of the professions and a hierarchical structure of social superiority with the architect and engineer at its pinnacle (Hindle and Muller 1996). The development of specialized professions led to the emergence of distinct occupational sub-cultures defined by unique beliefs, values, attitudes, languages, rituals, codes of conduct, codes of dress, expectations, norms, and practices. This emergence has not only damaged communications within construction projects, but has provided the foundations for the development of strong occupational stereotypes that have become deeply rooted into the modern construction industry's social fabric, influencing the way in which people behave toward each other. For example, Loosemore and Chin Chin (2000) found contractors were most often associated with negative stereotypes by other occupational groups while engineers were most favorably perceived, being characterized as proficient, systematic, confident, and composed. Managers have trouble eroding such deeply ingrained perceptions, particularly within the relatively short duration of construction projects. However, the most effective way is through procurement systems such as Design and Construct, which challenge traditional power balances, facilitate more participation, and require people to work together in multi-disciplinary teams.

Design management

Many construction problems arise from unresolved conflicts in design; therefore, better design management could contribute significantly to reducing the crisis-proneness of construction projects. For example, Bea's (1994) survey of engineering failures found that a majority had their origins before and during design, but that 98 percent were not detected until construction and operational phases. Indeed, there are numerous examples throughout history of disasters that have originated in poor design management. Perhaps the oldest documented example is the Quebec Bridge disaster in 1907. The bridge was to be the longest in the world with a span of 1,800 feet. However, disaster struck when, during construction, the structure collapsed, killing 82 construction workers. The Royal Commission of Inquiry's report, published in 1908, found that time and cost pressures had forced designers to compromise the designs and to "fast-track" construction on site with unfinished drawings that were only approximations of the final design. It was also found that during construction, designers had noticed two cantilever arms supporting 4,000 tons more than they were designed to support. Unfortunately, the problem was ignored so the imminent deadline for opening the bridge could be met.

In June 1995 a lack of design management was also a major contributory factor in the collapse of a shopping mall in South Korea that killed 501 people and injured 900. Investigations revealed that many illegal design changes were made, that designers were taking bribes and that they had not adequately monitored contractors on site. The majority of the design team, including the client and local government officials, were imprisoned for gross negligence. In response, many countries are now instigating legislation to make design teams more accountable for the long-term consequences of their actions. Nevertheless, legislation is no substitute for good management and with this kind of evidence, it is surprising that the construction industry has taken so long to focus on the problem of design management (Gray et al 1994). There is little doubt that the effective management of the design process can make a significant contribution to reducing a project's crisis-proneness.

Market problems

In contrast to the cultural problems described above, a harder economic position argues that the primary reason for the construction industry's crisis-proneness is its tendency to outgrow its market. Although one would expect industries and markets to naturally align themselves over time, construction companies have become survival experts from generations of exposure to stop-go government policies. The industry seldom has enough time to adapt before the next boom or bust reverses the trend again. This means the industry's size is invariably out of synchronization with its market and is generally larger. In such an environment, where too many companies are "chasing" too few jobs, the market forces margins and contingency allowances down and tempts companies to take risks that they cannot manage. This not only results in under-resourced projects but it creates conditions in which unscrupulous companies thrive. This confrontational environment encourages mediocrity, reducing the behavior of the industry to that of its lowest common denominator.

In the above environment, the clients' role is critical in that they determine the nature of the construction market. Unfortunately, apart from the most enlightened clients, which are relatively few in number, the majority have succumbed to the temptations of low price and have thereby perpetuated the problem. A good illustration of the dangers of this traditional cost-driven mentality, particularly on high-risk projects, was the 26 million pounds over-run on the 99 million pound Cardiff Millennium Stadium project, which was constructed to host the Rugby Union World Cup in Wales in October 1999. Many factors contributed to the cost escalations on this project, to the acrimonious disputes and to the serious delays that disastrously, could have forced a change of venue. However, a significant contributor was the guaranteed maximum-price contract that was signed with the main-contractor while the designs were still developing. The frequency of such problems in the construction industry has led to calls for more intelligent selection systems for contractors and consultants that consider a wider range of criteria than price alone (Hatush and Skitmore 1997). Furthermore, there has been wider advocacy of negotiated contracts and partnering arrangements that reduce the emphasis on price in the selection process (Latham 1994). Unfortunately, much of the industry has been slow to embrace these ideas and price remains the primary selection criteria for constructing project teams.

Man-made problems

Contracts

The cultural and economic conditions described above have nurtured the culture of conflict and mistrust that one associates with the construction industry. This is most vividly reflected in its voluminous, complex, and legalistic contracts which are underpinned by the following assumptions: that people's actions can be accurately controlled by those who have the legitimate authority of a contract behind them; that there is one best way to manage; that people cannot be trusted to do what is correct; that people prefer to be told what to do and that contract drafters know best. Much of the destructive conflict within construction projects emanates from the inappropriateness of this coercive contractual system. An emerging school of thought asserts that projects are better managed through a culture of mutual trust and collective responsibility. The underlying belief is that most conflicts are accidental and arise from misunderstandings between project members, and that such a culture can be created by simplifying the language and structure of contracts, making them less penal, more flexible, and more equitable. The Engineering and Construction Contract (1995) produced by the Institution of Civil Engineers in the UK, is based on these principles, although its effectiveness depends on eliminating the traditional divisive culture that traditional construction contracts have perpetuated. This has often proved difficult or has not been fully appreciated by managers, and the consequence has been a number of high-profile disputes.

Sub-contracting

The majority of disputes in the construction industry occur between principal contractors and sub-contractors. In fact, numerous studies have discovered a direct relationship between the incidence of sub-contracting and the performance of a project (NEDO 1983; Kumaraswamy 1996). Sub-contracting came about in response to increasing technological complexity and demands for faster construction times, but it also produced fragmentation, instability, short-termism, reduced customer orientation, legal complexities, unfair practices, and problems of communication, motivation and quality control.

The potential for disaster that can arise from sub-contractor management problems was well-illustrated in 1981 when two suspended walkways collapsed in the Kansas City Hyatt Regency Hotel, killing 113 residents and seriously injuring 186. Subsequent investigations by the National Bureau of Standards concluded that the collapse had been caused by undetected changes made by the steel fabricator to the original designs in order to make them more build-able (NBS 1982). This resulted in a serious reduction in load-bearing capacity to such an extent that the walkways could only just support their own weight, let alone any additional loads imposed by people.

Production problems

Uncertainty

Buildings vary infinitely in their scope and complexity and are produced in a relatively uncontrollable environment when compared to the products of many other industries. For example, how many managers in other industries have to take account of the breeding habits of bald eagles? This might seem ridiculous, but contractors involved in the construction of the U.S. highway through Snake Canyon in Wyoming had to stop work for six months because the site became a critical habitat for these birds. This meant that contractors had to work throughout the winter months which, in turn, necessitated special construction methods for sub-zero temperatures. While this particular event was predictable through the involvement of environmentalists, many natural events are not and it is not surprising that problems arise in construction projects.

People

Although attempts have been made to industrialize the construction process, the production of many engineering structures and in particular buildings is still essentially a craft-based, small-batch, out-of-doors process which, compared to most manufacturing processes, involves relatively little repetition, routine, or mechanization from one product to the next. In this sense, the construction industry is essentially a human one, and the process of managing construction, highly vulnerable to the unpredictability of peoples' idiosyncrasies. The vulnerability of construction projects to human error has been illustrated in numerous engineering disasters. For instance, we have already mentioned the collapse of the Quebec bridge in 1907, which the Royal Commissioners attributed to errors in judgment by the engineers who managed the project. Human error was also the underlying cause of other bridge collapses such as the over-ambitiously designed Tacoma Narrows suspension bridge in 1940; the Silver Bridge in Ohio in 1967, which killed 46 motorists; and the West Gate Bridge in Melbourne, Australia in 1970. In the West Gate Bridge disaster, the collapse was caused by engineers removing bolts to correct a misalignment at mid-span without appreciating the structural implications. Indeed, in a study of 604 construction failures in the United States between 1975 and 1986, Eldukair and Ayyub (1991) found that the majority were caused by insufficient knowledge, ignorance, carelessness, and negligence on the part of the engineer and contractor. Interestingly, while many of the contributory mistakes were technical in nature, 40 percent were managerial, relating to errors in work responsibilities and communications.

Project organization

The project organizations used to procure engineering structures and buildings have been referred to as "temporary multi-organizations" because they have defined start and finish dates and comprise people who are representatives of independent specialist organizations of a consultancy and contracting nature (Cherns and Bryant 1984). They are also highly transient in membership since the activities and specialists involved in the construction of a building vary over time. Furthermore, due to a traditional obsession with competition as a mechanism for selecting team members, it is common for teams to change entirely between different projects. These characteristics ensure that construction project organizations are made up of a constantly changing labor force that has loyalties to a wide range of interest groups. They also ensure that steep learning curves punctuate the continuously changing landscape of interpersonal relationships, particularly in the early phases of a construction project when the nature of the end product is often ill-defined or even unknown. The potential problems that can arise from these organizational characteristics were vividly illustrated in the Summerland leisure center fire disaster on August 2, 1973 in the United Kingdom that killed 50 people. According to Turner and Pigeon (1997), "a small architectural firm was undertaking its first large commission, designing a new kind of building, that posed new kinds of fire risks, and that was built with new kinds of construction materials. In addition, the conditions under which it was anticipated that the building would operate were changed significantly during the design process"(p. 46). This led to a catalogue of human errors, poor communications, misunderstandings, conflict and ignorance on the part of project members, all exacerbated by the time pressures under which the team was working.

MANAGERS AS A SOURCE OF CRISIS-PRONENESS

Despite having characteristics that would appear to be receptive to an open, flexible, people-orientated style of management, considerable evidence suggests that the managerial mind-set that predominates in the construction industry remains fundamentally scientific in nature. This is best reflected in the construction management literature, which indicates an increasing intoxication with popular business fads from mainstream management such as TQM, benchmarking, supply-chain management, and value-engineering.

Two recent buzzwords that have captured the imagination of managers in the construction industry are Business Process Re-engineering (BPR) and Lean Construction, which epitomize the increasingly radical approach to "*transforming*" organizations. While techniques such as BPR and Lean Production were designed to overcome the social problems associated with mass production by empowering workers to determine their own destinies, evidence indicates that their ruthless focus on productivity, process improvement, and efficiency, creates de-humanizing, under-resourced organizations that are more vulnerable to crises (Richardson 1996; Green 1998; Carmichael 1999). While re-engineering exercises need not involve down-sizing, in practice, they often do, and as Hall et al (1993) argue, they are often used as managerial facades to legitimize streamlining plans and to ask people to do more for less while paying token gesture to the introduction of meaningful change. The evidence to support this argument is compelling. For example, Xerox, one of the world's greatest advocates of re-engineering, spent $700 million over three years to shed 10,000 staff members. Furthermore, the Fortune 500 industrial companies in the United States "sweated off" 3.2 million jobs during the 1980s and four privatized companies in the United Kingdom (British Telecom, British Gas, British Airways, and Yorkshire Electricity) have recently re-engineered their operations to increase their collective turnover by 24 percent while reducing their workforce by 33 percent. This is worrying because one need look no further than the Westgate bridge collapse in Australia, the Ronan Point tower collapse in the United Kingdom, and the Challenger space disaster in America for examples of disasters that were contributed to by an element of stress from over-working. Indeed, not only does leanness increase crisis-proneness but it also reduces crisis-responsiveness by stripping away the spare capacity that organizations rely on, to respond to unexpected resourcing demands.

In addition to smaller and more pressurized workforces, new business trends have also produced a relentless shift toward more flexible employment practices.

Today, most organizations have fewer core employees than they did five years ago, and they rely on a workforce of corporate mercenaries who coldly drift from job to job with little sense of loyalty to anyone but themselves and with the prime objective of securing the greatest monetary return for their efforts (Loosemore 1999). This has created an environment of fear and selfishness that may well be contributing to the relatively high level of workplace accidents, discrimination, racism, and industrial relations disputes that characterise construction industries around the world.

HIGH-RELIABILITY ORGANIZATIONS

It seems ironic that the construction industry is attempting to create inflexible construction organizations precisely when the business environment is demanding more flexibility. According to Sagan (1993), a key feature of all "high-reliability" (low-crisis) organizations is redundancy and duplication. He illustrated this by referring to a number of situations where organizations have learnt to deal effectively with high risk environments. For example, U.S. aircraft carrier operations stress the critical importance of having both technical redundancy (backup computers, antennas, etc.) and personnel redundancy (spare people and overlap of responsibilities) in their systems. Overlapping responsibilities may seem inefficient in modern business terms, but on an aircraft carrier, it can be the difference between life and death by ensuring that potential problems that one person misses are detected by another. The same principles are used to manage nuclear power stations where independent outside power sources and several coolant loops, are incorporated into system designs, should existing provisions fail. Finally, Sagan illustrates that redundancy is also a feature of our body's immune system, which is why we can survive if certain body parts are severely damaged. For example, if one kidney is removed, the other can compensate. If our spleen is removed, our bone marrow takes over the job of producing red blood cells. These examples, illustrate that redundancy is essential to survival in a world full of potential risks.

THE ART OF CRISIS MANAGEMENT

While modern management innovations appear to help managers keep pace with a rapidly changing world, they have the opposite effect by reducing organizational flexibility, creativity and responsiveness. It would seem that the scientific school of management, masquerading as contemporary management trends, has little to offer the modern-day construction manager. Today's business world demands a different mind-set which sees management as an art rather than a calculated science, as it traditionally has been. This means less emphasis on controlling uncertainty via formal rules, prescriptive procedures, inflexible hierarchical structures, specialization, top-down information systems, close supervision, monetary rewards, and threats of punishment. Instead, uncertainty needs to be seen as opportunity rather than a threat, to be accommodated rather than suppressed. This is best achieved through flexible structures underpinned by a culture of openness, sensitivity, collective responsibility and trust. Today, people's idiosyncrasies and creativity are a vital organizational resource which harnessed, makes the difference between excellence and mediocrity and ultimately, success and failure.

CONCLUSION

Crises pose special managerial problems compared to day-to-day problems and the evidence presented in this chapter indicates that they will become an increasingly common aspect of managerial life. Unfortunately, traditional managerial values have diverted attention away from the need to build resilience into organizations to deal with the unexpected. To prepare better for crises, managers need to develop a mind-set that perceives them in a more positive way and that is more receptive to thoughtful, open, flexible, trusting and employee-centered management styles. Not only will this reduce the chance of catastrophic failure, it will release the untapped creative potential and energies that are yet to be exploited by traditional managerial practices.

However, while the employee-centered style of management appears more attractive, ethical, and appropriate for the future, this depends on the context in which a manager is operating. If the development of management thought has taught us nothing else, it is that *there is no one best way to manage in all circumstances*. Rather, the most appropriate approach to management depends on the nature of the task, the nature of the production technology, the instability of the environment, and the nature of the people being managed. In essence, a task-centered style is more suited to the performance of routine, uncreative and repetitive tasks undertaken in a stable, highly mechanized production environment by people with little desire for autonomy. In contrast, the employee-centered approach is more suitable for the performance of creative, non-routine tasks undertaken in a non-mechanized and unpredictable production environment by people who value autonomy. So, while there is strong justification for softening the scientific stance that dominates construction management practices, the most important determinant of success is being sensitive to context and responding accordingly. Unfortunately, those in charge of high-risk projects who need to make the greatest shift in mind-set are least likely to do so, because people in dangerous situations tend to seek safety in the familiar. In construction, that means numbers, rules, and procedures. This is one of the dilemmas of construction crisis management.

Chapter 2

Planning for Crises

Planning is the foundation of effective crisis management. This chapter discusses the various types of crises that can arise within construction and engineering projects and considers the appropriate response in each case.

INTRODUCTION

Preparation is essential in dealing effectively with a crisis, and the best-prepared organizations are those that have taken time to understand the different types of crises they may face. These organizations compile a crisis portfolio, prioritize it, and make considered decisions about those crises for which they should plan. However, they also constantly re-appraise the types of crises they are likely to encounter and if they fall outside existing categories, develop new strategies to deal with them. As Mitroff and Pearson (1993) found, a common cause of catastrophic failure in organizations is the temptation to focus only upon crises that are common to a particular industry or organization.

TYPES OF CONSTRUCTION CRISES

Several models of construction risks have been developed. A good example is Perry and Hayes' model (1985), which is reproduced in Table 2-1. While such models are useful in understanding risk exposure, they provide little insight into different types of construction crises and, in particular, into their managerial consequences. The focus is primarily upon causes. Consequently, there is currently little understanding of how to respond to different types of crises when they arise. However, mainstream crisis management research has developed crisis typologies that are based on both causes and consequences, and these are far more useful from a practical perspective.

MAJOR CRISIS CATEGORIES

Crisis management research has developed five broad categories of crises that are differentiated by their causes and consequences, namely: *technical; natural; political; social; and organizational. Technical crises* are defined by their human origins and their potential to cause major damage to human health and the environment. Outside construction, examples include the 1982 gas leak at the Union Carbide plant in Bhopal in India, which claimed 3,500 lives and caused

Table 2-1 A typical construction risk model (Source: Perry and Hayes 1985).

Category	Example
Physical	Loss or damage by fire, earthquake, flood, accident landslip.
Environmental	Ecological damage, pollution, waste treatment, public inquiry.
Design	New technology, innovative applications, reliability, safety. Detail, precision and appropriateness of specifications. Likelihood of change. Interaction of design with method of construction.
Logistics	Loss or damage in the transportation of materials and equipment. Availability of specialized resources – expertise, designers, contractors, suppliers, plant, scarce construction skills, materials. Access and communications. Organizational interfaces.
Financial	Availability of funds, adequacy of insurance. Adequacy of cash flow. Losses due to default of contractors, suppliers. Exchange rate fluctuations, inflation. Taxation.
Legal	Liability for acts of others, direct liabilities. Local law, legal differences between home country and home countries of suppliers, contractors, designers.
Political	Political risks in countries of owners and suppliers, contractors – war, revolution.
Construction	Feasibility of construction methods, safety. Industrial relations. Extent of change. Climate. Quality and availability of management and supervision.
Operational	Fluctuations in market demand for product or service. Maintenance needs. Fitness for purpose. Safety of operation.

more than 10,000 injuries (Shrivastava 1992). An example in the construction industry is the tower crane collapse which occurred on a London building site in May 2000. This killed three workers and could have injured many more, including passing pedestrians and motorists (Akilade 2000). Also included in this category would be the walkway collapse in the Kansas City Hyatt Regency Hotel (NBS 1982) and the West Gate Bridge collapse in Melbourne Australia (Bignell 1977). *Natural crises* have the same human consequences but differ fundamentally in that they are not man-made. Examples include the earthquakes which devastated Turkey in 1999 and the violent hail storm which hit Sydney, Australia in the same year, inflicting approximately $1.4 billion of damage to property (Clennell 1999). In contrast, *political crises* have their origins in political systems, wars, and public sector reform. For example, George (1991) analyzed a range of international conflicts including the Cuban Missile Crisis, the Arab-Israeli war, and the Gulf war, isolating the conditions that could precipitate an accidental war between the United States and the Soviet Union. In construction, a good example would be the costly and embarrassing seven year dispute over the UK's most expensive cladding contract at Portcullis House in London. This ended in 1999 when a curtain walling contractor successfully sued the House of Commons for selecting a more expensive bid at tender stage. Another example would be the public inquiry which threatened to delay the start of the new Wembley Stadium project in the UK. This arose out of a refusal by local government planners to recommend the scheme unless the developer made a 30 million pound contribution to a local development fund. *Social* crises relate to events such as the Rodney King riots in Los Angeles (Quarantelli 1993) and environmental protests such as those which caused considerable damage and delays on UK road projects in the early 1990s. Finally, there are *organizational* crises, which relate to high-profile corporate crises such as the Bearing's Bank scandal and Intel's Pentium chip flaw (Gonzalez and Pratt 1995; Sfiligoj 1997). In construction, a good example would be the public confidence crisis which Laing suffered as a result of its disastrous Millennium Stadium project in the UK. Also included in this category are labor relations crises like those which plagued the Jubilee Line extension in London, the first phase of which was delivered fourteen months late and more than 1.5 billion pounds over budget.

SPECIFIC CRISIS CATEGORIES

At a practical level, the broad categories identified above are of limited use because there is no indication whether, for example, technical crises demand a different response to organizational or natural crises. In a practical sense, more valuable typologies have been presented

Creeping, sudden, and periodic crises

Jarman and Kouzmin (1990) have distinguished between *creeping, periodic,* and *sudden* crises. These crises differ in their timing: a sudden crisis presents itself as one catastrophic event; a creeping crisis develops over time through a series of mutually reinforcing events that collectively escalate into a full-blown crisis. One example of a *sudden crisis* that has been all too common in the construction industry is the fatality of a worker. For example, while the European construction sector typically employs fewer than 10 percent of the working population, it accounts for more than 30 percent of all industrial workplace fatalities. In the year 1996-1997 alone, there were 90 fatalities on U.K. construction sites and non-fatal accidents were recorded at a rate of 15 per day (Anderson 1998). Indeed, in 2000, five years after the introduction of the European CDM regulations, fatalities were still at 70 per year and the number of major incidents had risen by 73 percent (Knutt 2000).

An example of a *creeping crisis* is continued sexual or racial harassment over a prolonged period of time that eventually results in a legal suit. In countries such as Singapore and Australia where the cultural diversity of the construction workforce is particularly great, discrimination has always been a serious problem which is badly managed (O'Rourke 1998). In the UK, similar problems exist, especially with increasing numbers of Eastern European immigrants entering the construction industry (Cavill 1999). Indeed, in the United States, equal opportunities suits have become so common that the Fortune 500 industrial companies spend over 2 percent of payroll on sexual harassment suits alone.

To broaden this typology, Booth (1993) has also pointed to the existence of an intermediate type of crisis that he has termed *periodic crises*. These occur at regular intervals but with varying predicability. Events that could cause periodic crises are annual budget cuts, changes in management, changes in government, etc. Booth also argues that organizations tend to respond to sudden, creeping, and periodic crises in different ways. For example, in response to a creeping crisis, organizations tend to increase their reliance on tried and tested procedures, rationalize it away or even ignore it. In contrast, periodic crises stimulate routinized responses manifested in rigid contingency plans that are constructed by internal bargaining processes between competing interest groups. Finally, sudden crises tend to produce a defensive response, particularly if no contingency plans were in place. After the initial shock wears off, which takes some time, a siege strategy can develop in which the organization becomes fragmented into its component interest groups. Booth's analysis is useful to managers in formulating an appropriate response because it allows them the foresight to put checks in place to mitigate potentially damaging behavior.

Perceptual and bizarre crises

Many crises are influenced by public perceptions, and these perceptions are, to a large extent, shaped by the mass media. Therefore, much hinges on the degree of understanding or, more often, the lack of understanding between industry and the media. Irvine (1997) produced a categorization system of "perceptual" and "bizarre" crises that relates to the role of the media in their creation. To Irvine, a perceptual crisis is a relatively insignificant problem that has been blown out of proportion by adverse media coverage. A bizarre crisis is one that has been fabricated by the media. Public scrutiny is becoming an increasingly important aspect of construction project management, particularly in landmark projects of national significance or in projects that have a major environmental impact. The issue of public relations is discussed in more detail later in this book but, to illustrate its importance, one can point to a whole range of construction projects in the United Kingdom that are perceived to be disasters as a result of what has been reported in the press. These include the new Royal Opera House, Channel Tunnel, Heathrow Terminal 5 (in its third year of public inquiry), the British Library, and the Channel Tunnel rail link project which, to the surprise of many, hasn't even started yet! Public relations are critical in such projects, because adverse media coverage can have an enormous impact on the morale and ultimately, the performance of project teams, resulting in a self-fulfilling prophecy for a hungry media. Perhaps, the greatest illustration of this is the Jubilee Line extension in London, which was a political instrument from the beginning, being announced at three Conservative Party Conferences. But then, as Gay (1998) states "it is a trophy project, so don't expect sympathy or the facts. Just keep your head down and dream of getting back to the simple life of simple contracts that simply make a modest profit, far from the public eye" (p. 29).

Triggering mechanisms

Benson (1988) categorized crises by their *trigger events* and Egelhoff and Senn (1992) indicated that these events could occur within an organization's *relevant* or *remote* environments. The difference between relevant and remote environments revolves around the directness of impact that events within them have on an organization. In essence, an organization's relevant environment comprises those influences that directly impinge upon an organization and represents a filter for events in the remote environment. Factors within the remote environment are divorced from the day-to-day activities of an organization and only influence it through elements in the relevant environment. For managers, this distinction is important because potentially problematic events within the remote environment are more difficult to detect and their impact is relatively indirect, creeping, and difficult to predict. An example would be the property crisis that developed in the late 1980s as a result of the stock market collapse. Over night this influenced the policies of financial institutions and reduced the viability of many projects, causing them to be shelved. In contrast, crises evolving in an organization's relevant environment are more easy to detect and relatively sudden in impact. A good example of a crisis arising from an organization's relevant environment is an industrial dispute revolving around pay or working conditions such as that which occurred in 1998 on the Jubilee Line underground extension in London, resulting in a strike by all 600 electricians on the project (Glackin and Barrie 1998).

Chains of events

Another useful way of distinguishing between different types of crises is to consider the chain of events that cause them. This strategy has been used for some time in safety science, based on the assumption that all accidents are precipitated by a chain of events traceable to its origins. Maps of such chains, which are often called "fault trees," may provide the basis for a greater understanding of crisis development and crisis management. To this end, one might argue that all crises are precipitated by a chain of events that can vary in three ways: *length, complexity,* and *conspicuousness.* Length refers to the number of events in a chain, complexity refers to the diversity of interdependencies between these events, and conspicuousness refers to their prominence and ease of detection. An example of a crisis with a long, complex, and inconspicuous chain would be a serious accident on a construction site that originates in the design phase by the incorporation of a dangerous detail or specification. The link between such an event and the eventual accident is so indirect and obscured by so many subsequent design decisions that it is likely to remain hidden until it manifests itself on site. However, the greatest challenges to managers are posed by crises that have short, inconspicuous, and complex causal chains because they have widespread implications, are difficult to detect, attack at a number of organizational levels, occur suddenly, and thus provide little opportunity for intervention. Furthermore, the complexity of causality is likely to create difficulties in constructing a response and to provide different interest groups with an opportunity to construct differing definitions of the crisis. In this respect, such crises have relatively high potential for conflict. In contrast, long, conspicuous and simple chains result in creeping crises, which are more obvious in definition, effect and present numerous opportunities for intervention and less potential for conflict.

CONCLUSION

Current frameworks of construction risk do not provide managers with knowledge of how to respond to different types of crises. This chapter has reviewed a number of classification systems from outside construction that might be useful in developing appropriate responses to crises. Different types of crises demand different responses and after identifing their organization's risks, managers can use this knowledge to plan appropriate strategies in advance. However, the typologies reviewed in this chapter provide only basic insights into appropriate responses to different types of construction crises and more research is needed in this area. Furthermore, by pigeonholing crises, typologies tend to oversimplify the complexity and uniqueness of every crisis. Standardized responses to crises may provide a useful initial response, but managers will need a detailed understanding of crisis dynamics if they are to deal with them effectively in the long-term. The following chapter considers this issue in more detail.

Chapter 3

The Dynamics of Crisis Management

This chapter divides crisis management into seven distinct phases: *detection, diagnosis, decisionmaking, implementation, feedback, recovery,* and *learning.* Each of these phases must be managed effectively if the crisis management process is to be successful. Unfortunately, during a crisis, people tend to behave in ways that make this difficult. The reasons for this destructive behavior are discussed.

INTRODUCTION

The purpose of any crisis management system is to maintain an alignment between organizational goals and performance by identifying and reacting to events that might cause major deviations. This system, which focuses on extreme risks, should be a specialized component of an overall risk management system, and its importance to an organization depends on several factors, the most obvious being the level of risk faced. However, all organizations are vulnerable to crises, no matter what the level of risk faced, and it is imperative that they should make some provisions for them.

DETECTION—THE FIRST PHASE OF CRISIS MANAGEMENT

The ability to react swiftly to any crisis depends on its early detection and this is achieved by monitoring potential risks to gather intelligence.

What is a risk?

Risk is exposure to the possibility of financial loss or gain, physical injury, damage, or delay as a consequence of the uncertainty associated with pursuing a particular course of action (Cooper and Chapman 1987). This definition is more useful than most because it emphasizes the opportunistic aspects of risk as well as its threatening aspects. Unfortunately, many definitions over-emphasize the latter, associating risk with exposure to peril or danger—thereby perpetuating people's anti-conflict values. Cooper and Chapman's definition is also useful because it draws a crucial distinction between uncertainty and risk. The distinction is that an uncertain event only becomes a risk for an organization when it has accepted the resourcing responsibility for loss or gain as a consequence of its occurrence. This is important because it indicates that the responsibility for good risk management practices lies in the hands of those that take, as well as those that transfer, risks. Good practice means not giving risks to parties who cannot control them, who do not have resources to bear them, and who have not been given the opportunity to provide a price for them (Abrahamson 1984). This mutual responsibility is under-emphasized in construction management literature and could be one of the reasons ineffective risk management practices continue to be a problem for the construction industry.

The breadth of risk in construction projects

The treatment of construction risks has been extensive in volume but narrow in scope. The main problem of the treatment is its focus, which is almost entirely upon *legal risks* imposed by legislation, common law, and the service and employment contracts that bind parties together. However, Gablentz (1972) also identifies those of a *moral* and *political* nature, indicating that an organization's risks are wider than its contractual responsibilities. Each of these types of responsibilities is discernible within construction project organizations. Moral responsibilities are imposed by the codes of conduct of professional institutions and society at large and political responsibilities are imposed by the culture of the construction industry and the organizational structure of individual projects. Morris (1998) provides a good example of this by describing the case of a developer who engaged an architect to design the shell-and-core of a shopping center. Contractually, the architect was not required to attend site during the construction period, saving the developer substantial fees. However, the architect was contractually required to supervise the fit-out, which the company had also been employed to design for the future tenant under a separate contract. In July 1993, fire destroyed the shopping center. Investigations revealed that construction defects had contributed to the spread of the fire. During court proceedings, it was held that there were no defects in the design that contributed to the fire but that the architect owed a general duty of care to the developer, beyond those in the contract, to report problems observed in construction of the shell-and-core while supervising the fit-out.

Another reason construction project participants have risks that are wider than their immediate contractual responsibilities is their interdependency. This was vividly illustrated in the Ramsgate disaster in the United Kingdom where a walkway collapsed, killing six people and seriously injuring seven. In this project, the client had relied entirely on a Design and Construct contract and, in accordance with it, did not interfere with the works. However, as a result of the accident, both contractor and client were convicted of a criminal offense, the former to the value of 1 million pounds, and the latter to the value of 400,000 pounds. This illustrates that despite the contract, the fortunes of construction project participants are inextricably linked and that it is folly to consider one's risks in isolation. The example also illustrates that organizations participating in construction projects are exposed to both voluntary and involuntary risks. This distinction has been neglected in the construction risk management literature, despite considerable evidence to suggest that due to unfair and unclear risk distribution practices, incompetence, ignorance, and time pressures, it is misunderstandings about involuntary risks that are the primary cause of conflict in the construction industry (Barnes 1991; Uff 1995). The problem with involuntary risks is that by definition, responsible parties may be unaware of them or reluctant to accept them.

Internal and external risks

Essentially, all risks exist either outside or inside an organization. This means that to be fully efficient, an organization needs both external and internal monitors.

External risk

Those who have external monitoring responsibilities for client risks should operate at the project boundary and "scan" the project environment for potential problems. Chapter Two discussed "remote" and "relevant" environments, leading to the conclusion that some risks are less obvious than others and less rapid in their impact. It follows, then, that to completely cover project risks, external monitors should have remote, relevant, long-term, and short-term dimensions to their activities. The importance of having a remote dimension to external monitoring activities is particularly acute in high-profile public sector projects where political maneuvering can have a significant impact on decisions at project level. Consider, for example, the Channel Tunnel Rail Link in the United Kingdom, which required a bill through parliament and the negotiation of Britain's largest planning application in history, which had to consider the opinions of approximately 258 pressure groups that were against the scheme. As Gay (1998) stated, "it did not need a project manager so much as a project resuscitator...." (p. 29). Similarly, Hemsley (1998) highlighted the importance of having a long-term dimension to external monitoring activities. He pointed to the problems that many companies suffered in the United Kingdom in the early 1990s as a result of not predicting the deepest recession in memory. To the majority of companies in the construction industry, the recession was a complete surprise and consequently, "all sorts of people found themselves in situations where their workload was reliant on a few market sectors, all of which were falling apart at the seams" (p. 33).

Internal risk

In contrast to external risks, internal risks tend to be far more "relevant," short-term, and largely the result of the industry's organizational, contractual, and employment practices manifesting themselves in the form of poor project performance. It is for this reason that internal monitoring activities focus on performance feedback that travels through management information systems (MIS), running vertically through an organization's structure.

Most organizations have an array of MISs, each specializing in information that specifically relates to one of the organization's range of goals. Most construction projects have five principal MISs that deal with cost, time, quality, scope, and functional information (Oberlender 1993). Each should be managed by a specialist with the necessary skills to interpret that information in a meaningful way. For example, in the United Kingdom, once site work begins, the quantity surveyor is primarily responsible for the cost MIS, the architect or engineer for the functional, scope, and quality MISs; the head contractor is responsible for the time MIS once work commences on site. Regardless of the professions that head them, the relative prominence of each specialist MIS should be a reflection of a client's goal priorities and the interdependency between construction project goals means that there should be cross fertilization of feedback between them. That is, the parties mentioned above should be constantly communicating and notifying each other of potential problems they discover. This is essential, because most crises have an impact across a wide spectrum of project goals. For example, the recent discovery of Aboriginal burial remains during a major engineering project in Melbourne, Australia, demanded foundation and design changes that affected its duration, costs, scope, function, and quality. For a complete response, it is critical that such events are fully communicated throughout an organization.

Monitoring problems

Unfortunately, while potential problems must be detected and communicated rapidly, monitors may be insensitive to the risks they are responsible for. If a potential crisis is not detected, it cannot be managed and managers must understand the reasons why this can happen.

A lack of focus

The most common reason organizations suffer a crisis is that no one realized the organization was vulnerable to it. This is best illustrated by moving outside construction to the Bearings Bank crisis where officials warned of possible calamity should Leesson's independent trading unit continue its current activities. Even after an internal audit in 1994 repeated these official misgivings, Bearings Bank sanctioned Leesson's double-role as chief trader and internal auditor, which enabled him to conduct his crippling activities unabated. Sheaffer et al (1998) discovered that the primary cause of this disastrous oversight was the confused state of the bank's boundary spanning activities and internal conflicts between different departments that had muddled lines of responsibility and accountability for the reporting of potentially damaging events. The huge time pressures under which people worked and the ability of Leesson to legitimately override the controls that were in place exacerbated this state of confusion, which resulted in risk not being monitored in some areas. Similar problems exist within construction projects where performance feedback is often deficient because of time pressures, conflicting interests, and arbitrary and unclear risk distribution within contracts.

An ignorance of risk management practices

Potential problems often go undetected in the construction and engineering industries because of the widespread ignorance of risk management techniques (Smith 1999). Risk management is crucial because it is about foreseeing potential pitfalls and identifying methods to avoid them and to deal with them in advance. The infinite variability of construction projects often makes this difficult. For example, on a recent wharf renovation project in Sydney, Australia, the highly unusual risk of workers contracting Hepatitis B from syringes washed into Sydney harbor was anticipated and minimized by establishing a regime of regular medical check-ups and hepatitis B injections for all construction workers.

Another problem is the perception that risk management is a construction phase process. Recent research in the US indicates that only 16 percent of designers consider worker safety, that 29 percent occasionally do so, that 45 percent never do so and that 10 percent might do so in the future (Gambatese 2000). This is despite considerable evidence that many project risks arise during design and can be identified and eliminated there. For example, engineers on the controversial residential building behind the Sydney Opera House, nicknamed "The Toaster", designed temporary steel lugs into the structure to receive safety barriers during construction. These were burnt off when the operation was complete. Other design measures to improve safety could include beam sizes that allow adequate headroom for workers, the specification of non-hazardous materials, component standardization to streamline the construction process and prefabrication to reduce material handling and storage on site.

Design generated risks are particularly likely when there are new or unusual materials, new or complex design guidelines and considerable cost and time pressures. An example of a project that included many of these elements was the Thames Barrier project in London which was constructed to avoid a repeat of the disastrous flood of 1953 that killed 300 people and flooded 64,750 hectares of farm land, 24,000 houses, 200 industrial premises, 320 kilometers of railway, 12 gasworks, and two electric power stations. Clearly, the potential environmental impact of such a project was huge, but the risk to construction activity was minimized by a sensitive design and by considerable efforts to involve the local community and environmentalists in the design process (Morris and Hough 1987). However, this project also illustrates the dangers of not adequately considering risks early enough in large engineering projects. While the barrier was undoubtedly a technical and environmental success, it must be classed as a failure from a project management perspective. The final cost of 440 million pounds was four times the original budget (although inflation was not allowed for) and the project took seven and one-half years to complete—almost twice as long as planned.

The reasons for failure were largely related to a lack of attention to identifying management risks associated with the project's implementation, which was in total contrast to the exhaustive investigations of technical risks in determining the project's viability. For example, there was little understanding of the effect that river level changes and river traffic would have upon the project program. Furthermore, the impact of deep-seated political tensions between local and central government, which affected the client's willingness to manage the project and exacerbated the already fraught industrial relations situation, were underestimated. Given the project's location and era, labor militancy was inevitable and more effort should have been made early on to secure a binding and comprehensive site agreement on pay and conditions.

In addition to analyzing risks early in a project's life, it also important to remember that the design process is influenced by a large number of people and that it never stops. This means that managers must be constantly vigilant to the influence of design changes on project risks. This was beautifully illustrated in the shipping industry in 1990 when Mitsubishi heavy industries had to recall its new generation of crude carriers because a large number of cracks were found in their cargo tanks. In subsequent investigations, it was found that, because the ship had been designed without consulting the builder, the ship construction yard unilaterally changed the design by, among other things, reducing the weight of the steel, thereby lowering construction cost, and reducing the time for construction. In addition to more continuous control, this financial disaster could have been avoided by involving all project stakeholders in the design process.

Competition between project goals

In addition to the problem of poorly focused monitoring, there are substantial difficulties in the cross-fertilization of feedback between different MISs on many construction projects. Baxendale (1991) found this to be a particular problem between MISs responsible for the supply of time and cost information because solving a cost problem often meant creating a time problem and vice-versa. As a consequence, problems detected in one system would not be communicated to the other, a deficiency that prevented a balanced and comprehensive crisis response.

The potential problems that can arise from competition among interdependent parts of an organization's control system was vividly illustrated on the 99 million pound Cardiff Millennium Stadium project that was being constructed to host the Rugby Union World Cup in Wales in 1999. According to Sir Martin Laing, the contractor's chairman, one of the reasons for the 26 million pounds loss suffered on that project was "a local baron" philosophy in their civil engineering division and a lack of management controls (Barlow 1999). One way of overcoming this problem is to share project risks, as was the case on London's Millennium Dome project, in which consultants equally shared responsibility for budget and time overruns. As David Trench, the site and structures director for the project, commented, this provided more rapid feedback, which gave one an instinct for when things were going wrong (Knutt 1998).

Changing project teams

Project managers face special problems in monitoring project risks because of their reliance upon the diligence of project teams, which are constantly changing. The dangers of changing team members within and between projects was vividly illustrated in 1991 when, during a ballast test operation, the Sleipner A oil platform sank 200 meters to the bottom of a fiord outside Stavanger, Norway. Accident investigators found that a contributory cause of this $1 billion accident was a loss of corporate memory associated with a complete replacement of the team who had been involved in similar platform projects where the problem had become well-established and simple solutions had been developed (Bea 1994).

A lack of loyalty

A further problem for project managers is the varied and often conflicting interests that characterize construction projects. This means that those monitoring client risks may not always have a client's interests in mind. To overcome this problem, construction contracts normally require project participants to monitor a range of specified client risks and react in a prescribed manner if they arise. There is little point quoting individual contract clauses here, but it is worth noting the obvious efforts being made in the Engineering and Construction Contract (ECC 1995) to overcome these potential problems. For example, the ECC gives consultants far broader monitoring responsibilities than they normally have, producing a far more integrated and wide ranging monitoring system. Furthermore, there is a requirement for "early warnings" if parties recognize the potential for a future problem. Such clauses are common in American construction contracts but not so in UK contracts, where clauses can often be construed to imply that only "actual" problems need reporting, providing a legitimate foundation for delayed notification and the escalation of simple problems into fully-blown crises. While early-warning requirements may be difficult to enforce, the American courts have recently demonstrated their willingness to do so. For example, in A.H.A. General Contracting Corporation v City of Houston, Court of Appeals of Texas, July 23 1998 (CE/02/M), a New York City Housing Authority awarded two construction projects worth a total of $4.7 million to A.H.A. Contracting with a requirement to warn of potential problems in a timely manner in order to receive the right to payment for extra work. However, A.H.A. waited until the end of the project to present claims for an extra $906,000 and the authority refused to pay anything other than minor extras. The Court of Appeals held that A.H.A.'s failure to notify of the problems in a timely manner prevented the owner from taking early steps to mitigate damages and precluded them from seeking recovery.

Cost and time pressures

It may be economically irrational or indeed impossible, due to limited resources, to monitor every potential risk an organization faces. For this reason, people tend to make "trade-offs" between the costs and benefits of doing so. For example, when the costs of sensitivity are considered too high, a conservative "lets think it over" predisposition develops, which results in warnings of crises lagging behind their occurrence.

This tension between diligence and costs burdens a project manager with heavy moral and economic dilemmas in establishing acceptable parameters for compromise. The moral dimension is particularly acute when considering safety issues because a price has to be placed on a person's health. The recent electricians dispute on the new extension to London's underground Jubilee Line was a good illustration of how the moral dimension can be forgotten under extreme time and cost pressures and how trade-off decisions in areas such as safety can be highly sensitive. While an array of problems led to this acrimonious dispute, electricians insisted that the main problem was managers' willingness to put workers at risk in order to increase productivity (Glackin and Barrie 1998). Glackin and Barrie documented a range of frightening safety lapses such as operatives working on a live 22kV transformer with aluminum ladders, an electrician being ordered to work on a live 415V panel before proper permits had been issued to isolate the power, operatives not receiving safety inductions and an operative having his finger amputated after losing his grip on a drill when another operative tripped over the power lead.

Defensiveness

Because monitors invariably generate information that threatens the status quo, organizations tend to develop norms that castigate such people (Argyris 1984). Furthermore, people are often required to provide higher standards of evidence than those who provide information that supports existing expectations and hypotheses. This is particularly so when intelligence information requires policymakers to do things they prefer not to do or that they are not prepared to deal with. The potential dangers of this defensiveness have been vividly illustrated in many disasters such as the Hillsborough football stadium disaster in the United Kingdom and the Challenger space shuttle disaster in the United States. In each case, chief decisionmakers, under pressure from external constituencies, filtered-out and ultimately ignored the advance signs of impending disaster (Jarman and Kouzmin 1990; Richardson 1993). For example, in the Challenger disaster, several aborted launches had led to ridicule from the press. This embarrassed policymakers and team leaders. On the night before the launch, engineers warned that the predicted temperature at the time of launch would be below the safety limit for the fuel tank seals. However, evidence indicates that engineers were pressured by their teammates to re-analyze their calculations about the fuel tank seals and subject them to stricter tests than would normally be required. This caused the engineers to question their original calculations, to place a greater degree of faith in the secondary seals than was normal, and to agree to the launch.

Superiority

Turner and Pigeon (1997) indicate that defensiveness can be a particular problem when the information that challenges existing mind-sets originates from sources outside an organization. For example, their analysis of the Aberfan coal slip disaster in South Wales in 1966, which killed 144 people, indicates that a sense of organizational exclusivity can lead to a sense of superiority over non-members. In Aberfan, the local council members foresaw the problem but were labeled as "cranks" and repeatedly "fobbed-off" with ambiguous and misleading statements such as "we are constantly checking these tips." As Turner and Pigeon (1997) argue, there was an "attitude that those in the organization knew better than outsiders about the hazards of the situation with which they were dealing" (p. 49).

In large construction projects, with annual turnovers that exceed many companies, defensiveness to outsiders can be a particular problem. The pressures, cohesion, loyalties, focus, and momentum that can develop on such projects can become so intense that they effectively seal themselves off from the outside world, considering outsiders as an unnecessary distraction and even covering up problems that may expose internal weaknesses to them. However, as Turner and Pigeon's example illustrates, the occasional involvement of outsiders, who are unfamiliar with a project is often the most effective means of detecting potential problems. Their exclusion only increases a project's crisis-proneness.

Shut down

Research in "killology," a new branch of behavioral science concerned with the psychology of killing in human combat, has shown that during a crisis, people tend to shut down all senses apart from the one that is most needed to survive under the circumstances (Bullock 1999). To illustrate this, Bullock cites an example of a playground shooting in the United States where two boys gunned down 15 teachers and students. In de-briefing sessions after the shooting, those directly affected recalled how their sense of sight and hearing had shut down according to their situation. After being initially stunned by the first shots, many people's eyes became their main sense—they began to see everything but heard almost nothing, just distant popping noises at the very most. In contrast, those that were actually shot seemed to go physically blind but hear everything that was going on in the finest detail.

The importance of this research for managers is that during a crisis, some monitoring systems may be shut down, leaving the organization vulnerable to further crises. This means that monitoring problems can arise because of crises elsewhere in an organization and that one of the challenges of crisis management is to prevent tunnel vision by maintaining vigilance to the early warning signs of further crises.

Timing

Monitoring problems may also arise because not all crises are equally visible. As Chapter One pointed out, some crises may be acute, sudden, and self-evident, while others may be creeping, emergent, and ambiguous. For example, some instances of poor monitoring may be related to the incremental manner in which creeping crises develop because this would make their early warning signs less easily detectable. This was the case in the Occidental Piper Alpha disaster on July 6, 1988. The long chain of events that led to the eventual fire were initiated by a simple unfinished maintenance job in the gas compression module (DOE 1990). A more bizarre example is McDonald's hot coffee crisis in which a customer was awarded $2.9 million in damages for third degree burns sustained from spilling a cup of coffee (Gonzalez and Pratt 1995). McDonald's records showed that prior to this highly public court case that subsequently led to numerous of copy-cat cases around the world, there had been 700 previous complaints about coffee burns.

Sudden crises may also be difficult to detect because managers tend to become flooded with a sea of contradictory information from which it is difficult to draw any sensible meaning. People have a limited capacity to handle information and are forced to "selectively perceive" incoming signals. Unfortunately, the natural tendency is for people to do so in a way that confirms their existing expectations and that suits their personal needs. Inevitably, the result is a self-fulfilling prophecy and the filtering-out of any early warning signs of pending crises (Comfort 1993).

Invincibility

Pascale (1991) highlighted the self-destructing nature of the world's largest and most successful companies. According to Pascale, "nothing fails like success" because a history of successes tends to induce a sense of invincibility and a "it can't happen to us" attitude that blinds organization members to potential problems. Furthermore, the growth that often accompanies success tends to confuse monitoring responsibilities and spurn bureaucratic systems that are slow in responding to potential problems that are detected. Indeed, this "back-lash" effect from past successes has been posed as one factor that contributed to the Challenger space shuttle disaster where the "can do" culture of NASA down-played the early warning signals of impending problems (Pearson et al 1997).

Mistrust, fear, and not speaking up

Chapter One identified the sources of mistrust and fear in construction projects. Ryan and Oestreich (1998) found that the most dangerous side-effect of such an environment is a hesitation to "speak up" for fear of punishment that creates a situation in which working relationships become clouded in *undiscussables*. These are secrets that everyone knows but which are only talked about privately. The danger is that the longer they remain undiscussed, the harder it becomes to talk about them and the more damage is done. Interestingly, Ryan and Oestreich found that certain types of problems were particularly likely to become undiscussables. These are summarized in Table 3-1.

Table 3-1 Undiscussable problems in organizations.

Category	Percentage of response
Management performance	49
Coworker performance	10
Compensation and benefits	6
Equal employment opportunities	6
Change	4
Personnel systems other than pay	4
Individual feelings	2
Performance feedback to respondent	2
Bad news	2
Conflicts	2
Personal problems	2
Suggestions for improvement	2
Others	9

The implications of creating an environment where people are afraid to speak up can be enormous. Ryan and Oestreich provide a good example of a financial crises in a small Midwestern software company that was a result of this phenomenon. Employees frequently described the CEO of this company as a tyrant. He created an environment of fear by bullying employees, publicly criticizing them, setting arbitrary deadlines without consultation, expecting unreasonable hours from everyone, and deliberately pitting people against each other. The negative climate resulted in the release of a faulty software package that cost the company $6 million, its reputation and customer base, and several key employees. Ironically, one employee had detected the fault earlier and voiced her concerns during a meeting. Her reward was her dismissal and a humiliating dressing-down in front of other employees. As one employee revealed, "The program was a disaster. The software wouldn't do what we said it would do. There was no ownership, no pride in the product. People were not asked for their input. More energy and time was spent covering your butt than on the quality of the program."

Shell Oil Company is a good example of an organization that has recognized that cultures of blame are unlikely to produce efficient or just outcomes. They have developed a management program designed to generate a corporate safety culture with no-blame practices. Interestingly, the accident record of Shell's tanker fleet, expressed in terms of frequency of injuries, has fallen dramatically since its introduction in the late 1970s (Horlick-Jones 1996).

The visibility of early warning signals

The visibility of early warning signals is another factor that is likely to determine the sensitivity of people to them. Essentially, the visibility of early warning signals is determined by their intensity, duration, and subtlety. Intensity refers to the strength of a signal; duration refers to the time over which it is detectable; and subtlety refers to its complexity and blatancy in terms of the degree of investigation required to detect it. Potential crises that emit low intensity, short duration, and subtle signals such as those originating in design, would be more difficult to detect than those that emit high intensity, long duration, and blatant signals such as those physically apparent in construction. The problem with mistakes made during design is that the high degree of reciprocal interdependency among design team members ensures that they are rapidly buried in and obscured by a labyrinth of subsequent and interdependent activities. In this sense, construction project managers must be particularly diligent to potential risks during the design phase of a project where monitoring traditionally has been lacking.

DIAGNOSIS—THE SECOND PHASE OF CRISIS MANAGEMENT

After detecting a potential problem, an organization should investigate it further to arrive at a diagnosis that will indicate an appropriate response. In control systems terminology the person responsible for this function is called the "comparator," and since the role demands specialist knowledge, such people should be professionally qualified specialists such as structural engineers, architects, quantity surveyors, cost engineers, or site managers. While the roles of monitor and comparator are distinct, ideally, the same person should perform them, because those who have collected specialist information are likely to be in the best position to assess its significance, highlight potential problems, and thereby ensure a rapid and appropriate response. By separating monitors and comparators, potential communication problems are introduced that can distort or slow down a crisis response.

Diagnostic problems

The diagnostic process is essentially one of data collection, and any problems encountered in the monitoring phase also affect the diagnosis process. However, some special problems are peculiar to the process of diagnosis.

Unclear goals

To make judgments about the potential impact of a problem upon project goals, comparators must have a clear idea of the standard of performance required in their particular specialist area of expertise. Unfortunately, as Kelly et al (1992) point out, even the most competent consultants can experience difficulties in identifying clear performance standards because of client inexperience, difficulties in identifying a client body, internal politics within client organizations, exclusion from early strategic reviews of client business processes, poor communication, and insufficient time for briefing. Further problems can be caused by continual changes in goals throughout the life of a project. Indeed, this was a problem that afflicted the new British Library in London, which exceeded its original budget by 400 percent and which provided spaces for only 12 million of the originally planned 25 million books. Here, changes in requirements were largely the result of changes in government policies and structural changes within government departments responsible for monitoring the project (Spring 1998). As the architect stated, "If you really want to devise an insane way of developing a building, you would go about it the way that the politicians have done. When Mrs. Thatcher came, it was constantly stop-go, stop-go, which was infinitely more expensive. They pulled the building up by the roots every 18 months, chopped a bit off, and put it back in the ground to grow." Comments from others involved in the project support this portrait of constant change: "They kept changing responsible departments, and every time the department changed so did the personnel"; "You can't run a long-term project on Treasury allocations that change each year and can't be passed onto the following year. This meant it had to be run on a cost reimbursable basis, so the consultants and contractors had no financial incentive to contain costs."

Indeed, these problems would seem to be common to many landmark public projects such as the $284 million flood control project in Las Vegas, which began in 1995 and will not be finished until 2007, six years later than originally planned (Feature 1999). This project, which is meant to protect Las Vegas from devastating floods has become something of a juggling act for engineers who must constantly chase federal funds and be ready to work whenever money becomes available.

Defensive routines

Further problems can arise in the diagnosis process because of "defensive routines"—behavior designed to prevent individuals or organizations from experiencing embarrassment or loss. This is particularly likely during a crisis because of the severity of potential ramifications and is even more certain if there is an element of personal blame associated with it (Argyris 1990). According to Sinclair and Haines (1993), uncertainty over patterns of responsibility for crises also causes this behavior because it is easier to justify ignoring a problem by passing it onto someone else. This is highly relevant to construction projects because many construction contracts are ambiguously drafted and there is a widespread inability to interpret them correctly.

Typical defensive behavior involves bypassing problems or covering them up by crafting messages that contain inconsistencies and discourage debate. Leavitt and Bahrami (1988) use the term "repression" to refer to this process of denying an unwelcome reality, and argue that it is a survival mechanism in the face of extreme change or threat. Unfortunately, this process creates an "organizational shadow," a darker side that haunts an organization until the repressed problems grow to the point where they have to be dealt with. This phenomena and its potentially disastrous consequences was vividly illustrated by Intel in 1994, who after discovering a flaw in its Pentium chip, declined to issue a recall or notify its customers. In a classic illustration of defensive behavior, Intel down played and trivialized the problem by arguing "no chip is ever perfect" (Gonzalez and Pratt 1995). Indeed, the company kept marketing the chip until IBM refused to use it in their computers and the value of its shares plummeted, causing a financial crisis within the company.

The research that led to the discovery of defensive routines was conducted in situations where the interests of those affected by a problem were similar. Consequently, the defensive behavior reported was largely uniform and widespread. However, people's behavior within a construction project is likely to be quite different because wide ranging and often conflicting interests will ensure that some party's interests will be served by covering up the problem, while others' will be served by its wider communication and even its exaggeration. This results in a wide range of tactics being employed by crisis stakeholders to force circumstances in their favor—one of the reasons construction crisis management is so challenging.

DECISION MAKING—THE THIRD PHASE OF CRISIS MANAGEMENT

If the diagnosis process indicates that the performance implications of a detected problem fall within acceptable tolerances, the crisis management process terminates at that point. However, if a comparator perceives an unacceptable threat, the process progresses to the next stage, which involves formulating an appropriate response. This is the responsibility of a person who has the authority to make the necessary decision to bring actual performance levels back into line with project goals. A decision can only be made by one person, even though it may be with the advice of many.

Decisionmaking problems

Problems of monitoring and diagnosis influence the efficiency of the decisionmaking process but numerous problems are also peculiar to the process.

Conflicting advice

Essentially, the decisionmaker's task is to choose from a range of alternative courses of action. These decisions may be self-generated, but are more likely to take the form of recommendations from specialist comparators who are more qualified to make such judgments. The challenge arises from the magnitude and complexity of crises, which means that they inevitably affect a whole range of project goals. This creates the need to synthesize potentially conflicting advice from a wide range of comparators—a process that demands a great deal of experience, sensitivity, political prowess, and a good understanding of the client's goal priorities. For example, in resolving a budgetary crisis in a construction project where costs are of the highest priority, an expensive design detail which adds to a building's appearance and to which the architect is attached, may have to be sacrificed to bring the project back within budget.

A further challenge for decisionmakers arises from resource constraints because decisions that exceed them will not receive corporate support from senior executives.

For example, there is little point approaching a client for extra money if their budget is fixed. While all clients have resource constraints, they may be particularly inflexible and acute in government funded projects in which issues of public accountability usually ensure that any increase in funding or extension of time will have to be audited before a decision is made. In the event of a crisis, there may not be enough time to wait for this process, and difficult trade-offs with more flexible and less conspicuous goals may need to be made. Perhaps the most famous example of an organization making trade-offs between different objectives is the Ford Pinto case (Haroon 1999). During the 1960s, the demand for sub-compacts was rising and the Pinto was to be the first in a new generation of lightweight, low-cost cars. The specifications were uncompromising and inflexible and the car was not to weigh an ounce over 2,000 pounds nor cost a cent over $2,000. To maximize sales, it was also to be in the showrooms faster than any other car in Ford's history. However, during design and production, tests revealed that in crashes over 25 miles-per-hour, the gas tank would rupture and cause fuel to spill onto the road. At 40 miles-per-hour, the doors of the car would jam shut and trap its occupants inside. However, production continued, Ford deciding to ignore the problem, calculating that the cost of compensating the estimated 360 deaths and serious burns per year would be about one-third the costs of recalling sold cars, stopping production, and redesigning the gas tank.

Centralized decisionmaking authority

Decisionmakers exists at all organizational levels and it is the significance of a threat that partly determines the level of decisionmaking invoked. Since crises are events of significant magnitude, they would normally command the attention of higher-level decisionmakers such as the project manager and occasionally the client. However, the level of decisionmaking invoked is also likely to depend on the degree to which decisionmaking authority is decentralized. A decentralized structure facilitates lower-level decisionmaking. This issue, of hierarchical location, is an important consideration in crisis management because the closer the decisionmaker to a threat, the faster the response, as the need for communication between organizational levels decreases. Highly centralized structures are notoriously slow to adapt to change. This often results in poor decisions that are based on the distorted and restricted judgments of one individual. The fastest response to a threat would arise when a monitor, comparator, and decisionmaker were the same person but; in the extremes of a crisis, senior people will want to be involved and it is highly probable that getting a decision will mean a tedious, frustrating, and potentially damaging journey through the organizational hierarchy.

The dangers of having a highly centralized management structure was illustrated by DeMichiel et al's (1982) analysis of accidents in the coal industry where it was found that low-accident mines allowed supervisors more freedom to make decisions and miners to suggest improvements that were genuinely considered and often implemented. Numerous corporate disasters over the years have illustrated this danger when an autocratic chief executive who controlled decisionmaking dominated companies. For example, the collapse of the Brent Walker empire in the 1980s was partly attributed to a single decision that the overwhelmingly persuasive Chairman, George Walker, made in buying the William Hill chain of betting shops for 689 million pounds. As Weyer (1994) points out, if there is one reliable indicator that a company will eventually head into trouble, it's having a charismatic, high-profile executive chairman who resists advice. No doubt, the same is true for construction projects.

Dealing with external constituencies

One characteristic of a crisis is that it affects high-level decisionmakers. However, the implications of the most high-profile crises are likely to burst through the top of an organization and implicate external stakeholders such as financiers, insurers, users, public pressure groups, unions, government departments, and others affected in the project such as sub-contractors and suppliers. At the very least, these external constituencies will need to be considered, informed of proposed solutions, and in extreme circumstances, handed the entire responsibility to deal with a crisis. For example, in the case of a site fatality, subsequent investigations by the health and safety inspectorate or union officials may deem that a site be closed until certain improvements are made. Similarly, in the case of bankruptcy, legally appointed company administrators, liquidators, or accountants may temporarily take responsibility for managing a project.

Dealing with the rights of external constituencies can be a complex, delicate, and highly political process with significant potential to exacerbate a crisis and delay response. Consequently, responsibilities for integrating these stakeholders into the decisionmaking process must be clearly defined in advance. The potential consequences of not considering the rights of external stakeholders were vividly illustrated in the Rugby Union World Cup Millennium Stadium project in Wales. During construction, poor relationships between the Welsh Rugby Union and its neighbor, Cardiff Rugby Football Club, led the latter to refuse permission for tower cranes to swing over its air space and for raked structural masts to over-sail its ground. This led to a very late redesign of the masts that supported the stadium roof and to the recalculation of loads on the entire structure.

Outside construction, a good example of the problems that external constituencies can bring was illustrated in the TWA flight 800 disaster, which killed 230 people. Subsequent analyses of this disaster have highlighted the problems experienced by crisis managers in deciding how best to deal with the crisis while considering the rights of distraught families, an eager press, an emotional public, and external emergency services (Fennelle 1996).

In the construction industry, the integration of external constituencies is likely to become a more prominent issue in the future. One likely source of increased intervention is the growing Green movement whose members have identified the construction industry as a major contributor to waste. For example, 25 percent of

America's solid waste is generated from construction activity and much of this waste, which can seriously damage the environment, can be either eliminated or recycled (Mincks 1996). Indeed, in high-profile road building projects around Europe, the civil engineering industry has already begun to experience the potential power of the environmental movement, facing considerable delays to programs and damaging publicity to the companies undertaking them. As new housing estates spring up around this infrastructure, a knock-on effect to the house-building sector is inevitable. Indeed, even those involved in the redevelopment of inner-city sites will be affected because, in addition to having to comply with increasingly stringent waste management requirements, the repopulation of many inner-city areas will demand greater respect for the quality of life of those who reside there. For example, in a recent 55-story commercial development in the center of Sydney, Australia, local residents had to be integrated into the management structure of the project and consulted about an extension of working hours needed to recover accumulating delays. Sydney is trying to repopulate its central business district and the power of the local residents was such that they were able to insist on a $250,000 bond that would be unconditionally sacrificed if the project generated any noise outside certain hours.

Irrationality

A decisionmaker's task is to make a decision and Brecher (1977) argues that the decisionmaker can follow a number of alternative routes. For example, when a decisionmaker has experienced a similar problem in the past, he or she tends to follow the cognitive routines that were developed for that situation. In other words, a decisionmaker will make an *automatic* decision by relying on a predetermined response. In contrast, when a problem is unusual, non-routine, and not anticipated in people's cognitive routines, they tend to follow an *analytical* process, assessing the costs and benefits of various alternatives. Finally, when a problem is unique, complex, surrounded by uncertainty, and difficult to measure in terms of its costs and benefits, Brecher advocates a purely *cognitive* model in which a person would make an intuitive decision based on gut-feelings.

Unfortunately, during the pressures and stresses of a crisis, decisionmakers are unlikely to follow the rational paths laid out within Brecher's framework. For example, Knight and McDaniel (1979) found that during a crisis, people tend to follow an inappropriate decisionmaking path by "process-following" in a situation that demands a non-routine, analytical response. People move into an automatic mode during a crisis for many reasons. For example, Knight and McDaniel suggest that the non-routine information a crisis generates is not easy to classify, involving lengthy and costly searches before it is understood. There simply may not be the time to do this information gathering during a crisis, particularly if attention has not been given to crisis management systems that can supply people with appropriate information at the appropriate time. Other researchers have shown that during a crisis, the ability to make an automatic decision may be an important survival mechanism (Bullock 1999). For example, studies of soldiers' physiology while under fire show that heartbeats can rise to 300 beats a minute, which causes a loss in complex motor skills, peripheral vision and hearing. For this reason, the army, in training people to kill, attempts to develop a conditioned reflex in which soldiers can move into automatic pilot and respond to combat in a predetermined way. In extreme crises, organizations need similar capabilities.

Reluctance to alter performance standards

A decision, whether it is *automatic, analytical,* or *intuitive*, should realign organizational performance and planned goals. With this goal, a decisionmaker may legitimately decide to *do nothing* if, for example, there is a possibility of the deviation correcting itself in time. However, in most instances, something will have to change, and logically, this can either be planned goals or existing performance levels. One other alternative is to attempt to reverse the force that is causing the deviation. For example, if a problem is being caused by a local government planner refusing to countenance a design change, it might be solved by reasoning or bargaining with him or appealing to an ombudsman.

Argyris (1984) argued that decisionmakers tend to shy away from the goal-changing option because "group-norms" can develop that castigate challenges to organizational objectives. One could imagine that such pressures would be strong in high-profile, landmark projects such as the Sydney Olympic Stadium or London's Millennium Dome where the consequences of altering budgets and programs would be subject to immense media criticism and scrutiny. For example, we can return to the Millennium Stadium in Wales that was being constructed for the start of the Rugby Union World Cup in October 1999. The Rugby World Cup draws the third highest TV audience of any regular sporting event in the world and the pride of a whole nation rested upon the successful completion of this project. Consequently, when the intense media scrutiny that constantly surrounded this project revealed it might not be completed on time, the contractor, John Laing PLC, was subjected to fierce public pressure. The potential humiliation to the Welsh nation of having to host the World Cup elsewhere was too unbearable to contemplate and the chairman, Sir Martin Laing, was forced to publicly reassure the Welsh public that "we are jolly well going to finish the damn thing" (Barlow 1999).

While the pressures to alter performance rather than goals may be high on major public projects, it may be the opposite on the majority of construction projects. Hermann (1963) argued that under the pressures of a crisis, standards that once appeared reasonable tend to appear unattainable and are likely to be relaxed because of the alleviating effect this produces. Indeed, numerous clients such as the British Property Federation (1983) have often complained that project managers too quickly advised them to lower their sights instead of demanding greater efforts to increase the project team's performance to stay on target.

IMPLEMENTATION—THE FOURTH PHASE OF CRISIS MANAGEMENT

By definition, crises have widespread *social, technical,* and *monetary* implications and any decision is likely to require a significant change in these aspects of an organization. *Technical* implications relate to modifications in the physical routines of the project; *social* implications relate to the way in which project members must alter their established patterns of relationships; and *monetary* implications relate to the extra resources required to implement those changes. In a highly interdependent organization like a construction project, such changes are brought about through a network of communications that originates from a decisionmaker and spreads throughout the organizational structure. For example, in a recent construction project in Melbourne, Australia, the unexpected discovery of Aboriginal burial remains required a design change in a building's super-structure which, in turn, affected structural engineers, designers, services engineers, contractors, sub-contractors, suppliers, and external interest groups such as the media, archaeologists, and Aboriginal interest groups.

Implementation problems

Problems implementing decisions are likely to arise for a number of reasons.

Poor communication

If implementation takes place through a network of communications between project members, then any communication problems will reduce the effectiveness of the crisis management process. We have already discussed the potential for communications to be damaged during a sudden crisis because of information overload and people's natural tendency to panic, become irrational, and shut down certain senses. However, aside from these natural phenomena, the nature of the construction industry can also contribute to communication breakdowns during a crisis. Communication difficulties have long been recognized to arise from historical divisions between construction professions; fragmentation of the process due to sub-contracting practices; sequential procurement systems that separate construction phases such as design and construction and conflicts of interests due to tendering and contractual practices. These problems create underlying tensions in all construction projects and it is during the pressures of a crisis that they are likely to be exacerbated and to surface by damaging the effectiveness of interactions among individuals.

Resistance to change

By definition, implementation involves technical, monetary, and social change, to which there is likely to be a natural level of resistance. People resist because all change requires the abandonment of past efforts and considerable rework. Furthermore, when risks are not shared, as is normally the case in construction contracts, change will inevitably create both winners and losers. Although winners are likely to support such change, Machiavelli reminds us that they are likely to provide only lukewarm support for change efforts, compared to the vigorous defense of the status-quo from potential losers.

External constituencies

The problems of integrating external constituencies into the crisis management process were considered in relation to the decisionmaking phase and they are mirrored during implementation. Indeed, here they can be more acute because the implementation process is likely to be more lengthy and widespread in its implications for organizational members, than the decisionmaking process. This was vividly demonstrated during the Zeebrugge ferry disaster in 1987 when emergency plans did not adequately account for the mass influx of the media who jammed communications lines to the extent that shore and sea-based rescue teams could not communicate with each other (Wagenaar 1996). A further unexpected problem was traffic chaos in the port area caused by an inquisitive public and the many emergency services needed at the scene.

In the temporary environment of construction projects, the integration of external constituencies is particularly difficult because there is little time to build a long-term working relationship with them. This is why it is unusual to find on a construction project, the kind of on-going relationship that most companies have with the emergency services such as the fire brigade. For example, mock fire drills are relatively rare on many construction projects, even though the risks of fire may be considerably higher at times. Another factor that causes difficulties in integrating external agencies into crisis management activities is the transient nature of construction project workforces. For example, imagine the difficulties of integrating into the crisis management process, sub-contractors who will soon be completing their work and moving to another project being funded by a different client and being organized by another contractor.

FEEDBACK—THE FIFTH PHASE OF CRISIS MANAGEMENT

Throughout the implementation phase, knowledge of results is critical to ensure that planned and actual performance are realigned. This information is known as "feedback" and it is acquired by internal monitoring of the organization to collect performance data. Performance feedback should be continuously assessed in relation to project goals and if necessary, further adjustments made until planned performance is in line with actual performance. This may take some time and in this respect the crisis-management process is likely to be a cyclical one that should continue until the organization is back on its feet.

Unfortunately, in the reality of a construction project, getting performance feedback can be a problem, due to factors related to monitoring, such as poor communication, conflicts of interests, time pressures, and difficulties identifying and measuring project goals. Acquiring feedback from those who stand to lose from a change is likely to be particularly problematic and an important aspect of crisis management must be the identification of potential losers. Even better, would be the creation of a risk-sharing structure that ensures such conflicts of interest do not exist in the first place.

RECOVERY—THE PENULTIMATE PHASE OF CRISIS MANAGEMENT

Depending on how well the crisis was handled, the aftermath may present significant psychological and physical damage that must be rectified and come to terms with. The initial emphasis during this phase should be on *recovery,* which involves returning an organization to "normal" as soon as possible. Recovery demands time and sensitivity since there may be internal and external investigations to handle and damaged relationships to repair. Furthermore, much of the psychological damage is not likely to surface immediately since people, traumatized by events, may bury unpleasant memories in their sub-conscious. In relieving these post-crisis traumas, the main problem for crisis managers arises from people's natural desire to look forward and from their reluctance to recall uncomfortable events. However, this issue must be fully addressed, not only for ethical reasons but because latent psychological disturbances may re-emerge at some point in the future and play a part in inducing further crises. As Bullock (1999) points out, the basic principle that underlies psychology is that people are only as sick as their secrets and that full recovery from a crisis depends on sharing them with others. The same is true for an organization.

LEARNING—THE FINAL PHASE OF CRISIS MANAGEMENT

Managers must not see the return to normality as the end of the crisis management process. As Gonzalez and Pratt (1995) point out, crises present profound *learning* opportunities by revealing important improvements for application to future crises. Furthermore, they can reveal weaknesses in an organization that would otherwise not be evident. In this sense, they can contribute to improved effectiveness through the *unlearning* of crisis causing behaviors and procedures that may be ingrained within an organization.

In many ways, learning and unlearning are arguably the most important phases of crisis management because much of the knowledge we use today to construct, manufacture, and operate engineering and built facilities has been acquired from analyses of past mistakes. For example, Carper (1989) points to the Gothic period of cathedral construction in Europe when repeated failures of spire towers such as that of the cathedral at Beauvais in France led to questions that extended the frontiers of builders' technical knowledge. In a more recent example, Pritzker (1989) cites a fire in a new hotel under construction in New York that led to changes in building codes and standards that now afford better fire protection to employees during construction.

Unfortunately, the time-pressured, temporary, transitionary, fragmented, and divided nature of construction project organizations does not encourage the far-sighted attitudes that inspire people to learn and unlearn. Evidence of this can be found in the emerging literature about facilities management that shows little evidence of the construction industry evaluating the effectiveness of its final product, let alone individual crises that occur during its production (Barrett 1995). Despite these potential problems, a desire to learn, however difficult and painful, is the key to preventing repeated mistakes and improving performance. This brings us to the role of the forensic manager—the person charged with investigating an incident to establish its lessons.

The process of forensic management

In addition to internal investigations to learn from an incident, a project may be involuntarily subjected to external forensic investigations with legal ramifications due to mandatory government investigations or claims for compensation by an injured party. Whatever the source of and motive for an investigation, everyone on a project is affected and it is useful to briefly consider what the process involves and how it is best managed to produce positive results.

Although every forensic investigation is different, they all involve data collection and analysis. Usually, an investigator works back from the time of an incident to its origin. The first step is to identify what may have changed between the time of the incident and the start of the investigation (Hendry 1989). For example, in the investigation of a fire, rescue services may have disturbed the scene by cutting fallen beams to release trapped workers. The second step is to determine and map the exact sequence of events that led to the incident by using techniques such as fault-tree analysis. For example, if a chemical spill on a construction site caused toxic fumes and, as a result, made workers sick, where was where was the initial point of discharge? How were the chemicals released? How much chemical was discharged? Where did the chemical flow? How was the chemical contained? What emergency procedures were in place? Were safety procedures followed? The final phase of analysis is the evaluation of pre-existing conditions that set the accident in motion. For example, was a vital piece of equipment missing? Was there adequate training for employees? Were employees "larking around"?

Forensic investigation data is usually collected via interviews, observation, surveys, photography, and document inspection. Often, inspections of documents and interviews with key personnel may need to be repeated. In this sense, full access to the site, to its personnel, and to its documentary records must be afforded to the investigation team if the process is to be effective. The process may also involve laboratory work to test materials or even a full simulation or reconstruction of events surrounding an incident. Either of these options could mean temporary closure of the site.

Clearly, the approach to data collection depends on the nature of the event being investigated. For example, structural failures such as the collapse of the Kansas City Hyatt walkway in 1981 may require significant testing of structural members. In contrast, during a recent tax avoidance crisis on a $400 project in Sydney, Australia, the focus of investigations was wholly upon people. Subsequent investigations by tax officials resulted in all of the sub-contractor's employees being interviewed and served with demands for $30,000 tax bills, which caused major disruption to morale on site and to the physical progress of the work. Further investigations by union officials and by managers from the principal contractor eventually resolved the problem, but not without considerable disruption.

Clearly, investigations need to be managed sensitively to avoid severe disruption, in a physical and psychological sense, to a project. One way to do this is to ensure that everyone knows what the investigative process will involve and if necessary, is given support to get them through it. Psychological support is particularly important if the process may involve recalling very unpleasant memories or revealing evidence that may implicate working colleagues.

The forensic team

Since most crises will have a variety of technical, physical, procedural, and human causes, the forensic manager would normally coordinate a team of people who have the necessary range of skills to investigate these dimensions. This may include such divergent professionals as management consultants, psychologists, police, fire investigators, engineers of all types, union officials, lawyers, traffic accident investigators, accountants, manufacturers, surveyors, and pure scientists. One of the dangers of forensic investigation is the tendency for various interest groups to direct its focus. This is because most investigations take place within highly political environments and may have important financial implications for people and organizations in terms of legal actions and insurance payouts. Parties with considerable financial clout can be particularly persuasive in such situations, as can individuals who are personally injured by an incident and who need some form of compensation to help rebuild their life. However, an investigator must not be swayed by personal feelings, sympathies, and external pressures because the impartiality of the process and people's trust in it is essential for it to have any value. For example, Hendry (1989) describes an accident in which commercial pressure, personal sympathies, and distorted data from injured parties could have easily interfered with the investigative process. In this accident, someone who was leaning against a tower crane when it touched an overhead cable suffered a severe electrical injury. The success of the injured party's insurance claim depended on his proving that the cab was badly manufactured in terms of the visibility it afforded to the cab driver, and that the cab driver and crane director on the ground were not negligent. However, after inspecting excellent site photos of the accident, examining the crane, and talking to those involved, the investigation concluded that there was no contributory deficiency on the part of the crane manufacturer.

To ensure impartiality in the investigation process and widespread acceptance of its results, it is often best to appoint an "outsider" who has the respect of all parties to manage such investigations. Expert representatives of all potential stakeholders who may be implicated in blame should also be included. Such a complex team of divergent skills and interests clearly needs skilful management, and this task is made easier if the forensic manager is competent, is able to make difficult decisions, is able to offer a timely opinion, and is able to perform an objective analysis of the various pieces of, often conflicting information, which he or she will receive. Furthermore, it would help if members of the team are not only experts in their field, but that they have prior experience of such investigations, enjoy field work, are trustworthy and ethical, are determined, are able to deal with ambiguity and frustration, are good communicators, and have interminable patience. There is no substitute for these attributes since no amount of expertise in an area will prepare an investigator for the physical and mental chaos that surround the scene of most crises.

The forensic report

Clearly, the whole process of conducting a rigorous investigation takes considerable time and money and it should be started as soon as possible after the event, as evidence is often destroyed during rescue or clean up operations. Furthermore, sabotage by guilty parties may result in the destruction of important documents; important witnesses may leave a project; and the memory of potential witnesses may fade. The end result of the process is a report that should identify the causes of an accident rather than to allocate blame. However, investigators are often asked to highlight illegal or improper activity or to produce a damage evaluation. Furthermore, it is inevitable that some forensic reports do implicate people in blame and consequently, it is useful if the forensic manager has some knowledge of legal processes. While implications of blame may be unavoidable, the investigative process must not degenerate into a "witch-hunt" for a scapegoat, which is a common reaction to the powerful psychological shock of a crisis (Horlick-Jones 1996). Although the identification and punishment of wrongdoers has some potential to prevent recurrence, blaming others can easily backfire and detract attention from the important lessons to be learned from a disaster. This is what occurred in the Exxon Valdez oil spill when, during the post-crisis phase, Exxon sought to avoid blame by focusing their energies upon discrediting the coast guard rather than seeking to learn from the disaster. Not only did this prevent Exxon from learning many important lessons, it led to further public relations problems that merely exacerbated the costs of the original crisis. Blamism also occurred in the aftermath of the Purley rail crash on March 4, 1989 in London, where 5 passengers were killed and 88 were seriously injured (Horlick-Jones 1996). After this disaster, government investigations blamed one of the drivers of the train. As a result, he was prosecuted for manslaughter and received an 18- month jail term. The report and courts dismissed behavioral research that showed that the repetitive tasks involved in monitoring warning signals can produce a mind-set in which a person could believe that he or she had performed a task when, in fact, that was not the case. The decision caused an outcry with rail unions threatening strike action and claiming that there was one law for managers and another for workers.

While the best forensic reports stick objectively to the facts, sometimes it is simply impossible to determine the causes of a crisis. In such circumstances a report should not indulge itself in speculations because the forensic manager is responsible for protecting the professional reputation of all parties under investigation at all times. While many forensic teams carry out their work under the specter of legal action, few reach court apart from those that involve injury or loss of life. For this reason, reports should be written more constructively, as a basis for constructive negotiation between potential stakeholders implicated in a crisis. In this sense, such reports, if undertaken with the cooperation of all parties can be a useful method of third-party dispute resolution.

CONCLUSION

This chapter has identified the phases of the crisis management process that must be managed efficiently for a reaction to be successful. These phases are detection, diagnosis, decisionmaking, implementation, recovery, and learning, and their interrelationships are illustrated in Figure 3-1. This model isolates a series of distinct but interrelated phases of activity, each of which must be performed effectively if the overall crisis management process is to be effective. The model also identifies the roles of monitoring, diagnosing, and decisionmaking, which people play in the crisis management process. These functions can be performed by separate people or combined under the responsibility of one person, and it is likely that the efficiency of crisis management would, in part, be determined by the extent to which these responsibilities are separated. Any problems in the performance of these functions would likely lead to inefficiencies in crisis management. Indeed, crises seem to have an unfortunate tendency to create conditions that would be conducive to problems, which suggests that inefficiencies would be probable rather than possible.

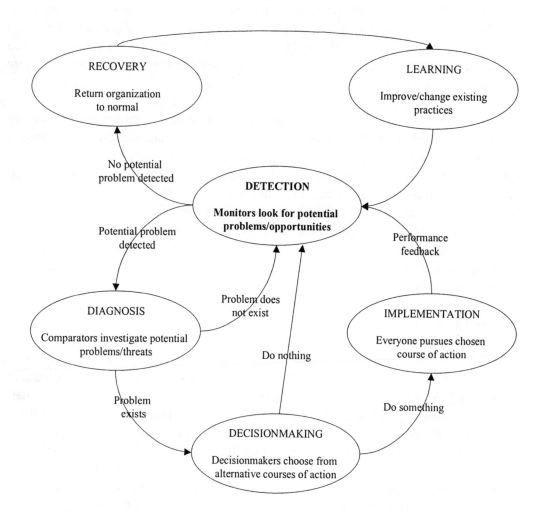

Figure 3-1 A model of crisis management.

Chapter 4

Emergency Planning in Construction Projects

Many of the problems discussed in the previous chapter can be attributed to a lack of preparation. Planning is the "bedrock" of effective crisis management because it helps avoid the damaging chaos that characterizes the early stages of a crisis.

INTRODUCTION

The importance of a well-conceived crisis management plan cannot be overstated. This has been illustrated many times, such as in the Occidental Piper Alpha disaster where appropriate operating manuals on how to interrupt a potentially catastrophic sequence of events were almost totally lacking (Bea 1994).

Unfortunately, all too often, organizations think they are prepared for crises because they have reviewed insurance policies, implemented back-up computer systems, and installed sprinkler systems. The fact is, the average insurance pay-out only covers one-eighth of the financial costs of an insured event and insurance policies cannot compensate for the indirect costs of a crisis arising from damaged employee morale and tarnished customer relations. To illustrate this, we return to the much-maligned Ford Pinto and Ford's calculated decision to rely on insurance pay-outs to cover the costs of potential legal suits. This proved a disastrously misguided decision because Ford had to recall 1.5 million cars and pay out compensation claims that were more than three times what was anticipated. Furthermore, they became the first American corporation to be indicted on criminal homicide charges and in 1980, almost 10 years after the car's production, Ford's reputation for safety was so bad that it ceased production of the model (Haroon 1999).

EMERGENCY PLANNING

The subject of emergency planning is receiving considerable attention in the crisis management literature, although many companies are reluctant to undertake the process (Mitroff and Pearson 1993). Indeed, in the construction industry, Teo (1998) found that corporate philosophies do not support crisis management and that any emergency planning is restricted to safety issues. Planning for other types of crises is almost non-existent and when it is present, is undertaken in an insular, informal and haphazard manner and supported by little strategic guidance and resources. One might expect this in an industry where over 90 percent of firms employ fewer than 10 people. However, Teo's research focussed on the world's largest contractors involved in the world's most complex projects and it would be reasonable to have expected some consideration of emergency planning.

Not surprisingly, the most crisis-prepared companies in the world are those operating within relatively high-risk industries such as shipping nuclear power, chemical processing, airport management, and oil. For example, Shell Oil uses crisis planning as part of their long-range strategic planning process and find that it often leads to many unexpected and indirect benefits such as new ideas that produce more innovative ways of doing business. Shell also finds that crisis planning highlights interdependencies and weaknesses that may otherwise go unnoticed and can bring people closer together by forcing them to act out and think through extreme scenarios. The emergency services have also invested a considerable amount of time researching and advising upon such plans and the construction industry could benefit from their collective experience. This is particularly true for companies involved in high-risk projects of national and international significance that are subject to public scrutiny, have a relatively high potential for large numbers of causalities, and have the scope to involve many external agencies.

THE AIMS OF AN EMERGENCY PLAN

Emergency plans should be constructed prior to a project's commencement and cover major risks in all stages of a construction project from inception to operation. Emergency plans are designed to mitigate the potential damage exacted by crises by enabling the most rapid response. They can do this because they are constructed outside the pressures normally responsible for the destructive behaviors which tend to develop during a crisis and magnify its impact. Having a preconceived plan that can be automatically implemented takes away some of the initial pressure and shock associated with the early phases of a crisis. This creates a valuable "breathing-space" within which people can calmly investigate the problem and agree on an appropriate response. The importance of a good start in crisis management cannot be over-stated. During a crisis, every second counts and the first few hours are particularly critical. This is especially true if external constituencies are involved because initial impressions play a disproportionately large role in shaping their judgments of competence and blame. If initial impressions are bad then an organization will be judged guilty until proven innocent and in many instances this can intensify a crisis and accelerate its escalation. This was demonstrated when John Laing PLC issued a profit warning as a result of its severe losses on the Rugby World Cup Millennium Stadium in Wales. As Sir Martin Laing stated, (Barlow 1999) "As soon as we issued the profit warning it was obvious that everyone in the construction press was going to be interested…. But there are a few twists that could have been put on it that are a little more positive than those that came out" (p. 24).

PREPARING AN EMERGENCY PLAN

Many organizations in high-risk industries have a permanent disaster committee that is responsible for championing the need for crisis management, identifying current preparedness and vulnerabilities, devising disaster plans, and coordinating people during a crisis (Kutner 1996). The membership of such committees is an important factor in determining their ability to do this, and they should consist of senior managers, managers from all functional departments, and external professionals who have experience of crisis management, public relations, the law, and physical and mental health issues. In particular, commitment from the top of an organization is essential if the activities of a disaster committee are to be taken seriously and if they are to have a chance of success. The various aspects of these activities are discussed below.

Conducting crisis audits and creating crisis portfolios

A crisis audit assesses an organization's crisis capabilities and identifies the inherent risk factors in its environment, internal activities, technology, infrastructure, and culture that need to be addressed to improve its crisis preparedness (Mitroff and Pearson 1993). The first stage in this process should be to develop a working definition of a crisis from the organization's viewpoint and to then identify and rank, in probability and consequence terms, the *types* of crises the organization is vulnerable to. This involves learning from past events, looking into the future and exploring unusual combinations of events that may seem unlikely, but could combine to produce a serious crisis. Ranking allows appropriate judgments to be made about the relative costs and benefits of constructing a crisis management plan in each case because planning for every possible crisis is not economically rational.

The importance and difficulty of exploring an organization's crisis vulnerabilities was well illustrated in the debilitating crisis that the Australian insurance company, Manchester Unity, suffered in 1993 (Forman 1993). This crisis originated from routine maintenance on its headquarters that caused a fire and, in turn, destroyed the power supply for the whole building. Since insurance companies are almost completely computer reliant, the lifeblood of the business was cut off in seconds. While Manchester Unity had embarked on emergency planning, its plans dealt only with extremes such as bomb scares and serious fires. No one had thought what might happen if the building was still standing and functional without any of the modern means of communication that were central to its business. Consequently, maintaining business operations were difficult and eventually a decision was made to set up a temporary communications center in a local hotel and to transfer head-office staff to suburban branches. Although the crisis was handled well, there is little doubt that it could have been handled better if these events had been anticipated and if simple measures such as manual typewriters and filing systems had been available as a back-up system.

Establishing monitoring systems and standard operating procedures

One aspect of the disaster committee's job is to establish monitoring systems to detect potential crises. The disaster committee should also develop standard procedures that define precisely *who* should be involved in a crisis response, *what* they should be doing, *when* they should be doing it, and *how* they should be doing it. These procedures, in effect, establish a pre-defined emergency communication network that needs to be followed during a crisis' early but critical phases, when people are disorientated by events. The intention is to "buy" the organization some time to come to terms with events, to allow people to re-orientate themselves, and to ensure that appropriate resources are mobilized quickly and that they are commensurate with a crisis' scale. To do this, the procedures should be achievable, simple, flexible, and understandable by all internal and external stakeholders. For example, in Australia and Singapore, construction sites have many migrant workers and this may require the production of manuals in a range of different languages.

In most crisis-prepared organizations, emergency procedures are set out in clear and easy-to-follow manuals with which everyone is familiar. Normally, such procedures consist of a number of tiers, the first being instigated during the very first hours following a crisis when there is little information available about it. This lack of information demands that these initial procedures should be as standardized as possible, covering broad families of crises. For example, many types of crises may require an evacuation of the site and others may have common health implications that demand first-aid treatment. Grouping crises into specific categories enables emergency procedures to be simplified, ensures a more rapid response, and thereby minimizes the chances of the incident escalating. However, the effectiveness of such generic procedures is limited, since different types of crises quickly demand different responses. Consequently, when the type of crisis has been identified but a detailed response not yet formulated, a second and more detailed tier of procedures should be initiated. For example, fire authorities often employ a graduated response, ranking accidents as Category 1,2,3, or 4 depending on their seriousness. A range of emergency plans is then formulated to match the weight of each category in terms of the people, time, and resources involved (Davis 1995). Once again, such procedures are only designed as a temporary response to alleviate pressures upon the project team and eventually, to buy time for the development of a strategy that is specifically tailored to the unique demands of the crisis at hand. An over-reliance on standard procedures will almost certainly lead to an ineffective crisis response.

Creating a command center

During a crisis, information is constantly being generated from a multitude of sources and it is critical that it is supplied "live" to the correct place, at the correct time, and in an understandable format (Davis 1995). In this sense, a key aspect of a disaster committee's job during a crisis is to identify a clear command center that represents a single point of responsibility for decisionmaking and information management. Such centres are a critical coordination mechanism that helps facilitate a unified crisis management effort since one of the greatest problems that can emerge during a crisis is the tendency for people to act independently. For example, in the case of a fire emergency, the command centre should have the sole responsibility to contact emergency services and to coordinate individual supervisors who are charged with clearing certain areas of the site. In the case of an economic crisis such as the bankruptcy of a major sub-contractor, the command center should be responsible for reorganizing work and re-employing another sub-contractor. In addition to being of practical importance during a crisis, command centers also play an important symbolic role. Nicodemus (1997) provides an example of a company that faced a crisis and named their command center "the war room," where they declared war on the problem.

Security

Security is another important issue for a disaster committee to consider since interference from unwanted elements can exacerbate a crisis or, at the very least, interfere with its management. This involves identifying external constituencies who feel that they have a stake in a crisis' outcome but who cannot contribute to its solution. Those involved in crisis management efforts should be insulated from these disruptive elements so they can develop a strong focus on the problem.

In some situations, it is also important that the site of a crisis is physically cut-off from these elements, particularly when it continues to represent a danger to the public. In such situations, evacuation procedures may need implementing and it is essential that they are clearly communicated to everyone on a project and reinforced by regular training and mock-drills. For example, public address systems, sirens and horns can be used to notify people of an incident if they are placed at strategic locations so everyone can hear them. Whatever signal is used, it must be as simple and as unequivocal as possible. Responsibilities for using them must be clear, as should appropriate back up if, as Murphy's Law dictates, key people are away on the day of an incident or if essential equipment malfunctions. An important part of evacuation is the clear labelling of exit routes from all parts of the site. In particular, people should know that mechanical hoists cannot not be used in an evacuation and that all potentially dangerous machinery in the vicinity of escape routes must be switched off. Since a construction site is a constantly changing physical environment, the positions of notices and their maintenance needs constant monitoring. Furthermore, all evacuation routes should follow the shortest possible route to checkpoints where role-calls can be taken in safety. They should also be wide enough to facilitate an orderly evacuation of the building. On inner-city sites, this may be the street and the hazards to the public, to traffic, and to site workers must be assessed in association with public services such as the police.

The potential danger of not having adequately thought out evacuation plans and well-marked evacuation routes was demonstrated in the Beverly Hills Supper Club fire in May 1977 which killed 164 people. The official investigation report reveals that the club had no evacuation plan and that employees were not schooled or drilled in the duties they were to perform in the case of fire. Furthermore, means of egress were not marked and the escape route itself was too narrow to take the number of people who were in the building at that particular point in time (Best 1977).

Developing a culture of collective responsibility

The need to insulate a disaster response team from unwanted elements does not mean it should be allowed to become introverted. Consideration also needs to be given to the reorganization of non-crisis management activities so the remainder of an organization can function as normally as possible. Crises inevitably drain a considerable amount of energy from other functional areas within an organization, demanding special efforts from the people who operate there. Clearly, without a considerable degree of peripheral goodwill and a sense of collective responsibility, the impact of a crisis can spread to other parts of an organization. Such goodwill cannot be expected if it did not exist before a crisis, and in this sense, the crisis management process needs to be continuous.

One way of developing a culture of collective responsibility is to communicate everyone's interdependency during a crisis and to clarify and, ideally, share project risks as much as possible. Most crises demand an injection of extra resources into a project and if the disaster committee does not identify their source in advance, then a crisis will stimulate negotiations and potential conflicts that will delay a response.

Decisions concerning risk distribution are particularly relevant to economic crises, and earlier we provided evidence to suggest that they have been a major cause of conflict within construction projects. We also identified a series of principles to guide risk decisionmaking. These principles apply at all points along the contractual chain and to consultants, as well as contractors. It is also important to realize that the client's initial risk management practices are inevitably transferred along the contractual chain. For example, if a contractor is employed under a high-risk contract and has not been given the opportunity to price for those risks, then it is likely that they will attempt to transfer those risks along the chain by using back-to-back contracts and similar employment practices with their sub-contractors. Indeed, sub-contractors may do the same and so on, until all project risks have been dissipated to the end of the contractual chain. Unfortunately, it is here that the most vulnerable, crisis-prone organizations exist, and when problems begin to occur that demand extra resources, the end result of this risk-cascade is inevitably a backlash of conflict up the contractual chain as parties deny any responsibility for them.

Public relations

Public relations are an essential aspect of crisis management since most types of crises have implications beyond an organization's boundaries. In essence, the three "publics" that need to be involved in a crisis are employees not directly affected by it, external and quasi-external interest groups and the general public. We have already discussed the first two publics and it would be foolish to ignore the third. As Aspery (1993) argues, "crisis communications built on well-established relationships with key audiences stand a better chance of protecting, even enhancing your reputation during difficult times. A company which decides to start communicating during a crisis will have little credibility" (p. 18).

Unfortunately, construction companies tend to attach little importance to the building of sound relationships with the media, seeing it as a non value-adding activity and perceiving most journalists as dangerous, untrustworthy, and irresponsible (Moodely and Preece 1996). This rejection of the media tends to be particularly strong during a crisis when organizations look inward and consciously hide from the public, seeing them as an unnecessary distraction to rescue efforts. However, this is precisely the time when it is most dangerous to ignore the media, since in the aftermath of a crisis, the public has a tendency to embark on a process of ritual damnation. This is particularly true of high-profile, publicly financed projects in which people may feel a greater right to recrimination as a result of having paid their taxes to finance it. As Horlick-Jones (1996) notes, "Since the abolition of capital punishment, the British public has turned to those in charge during lurid disasters to satisfy its lust for retribution. Find someone to blame, cries the mob, and off runs Whitehall to offer up someone for lynching" (p. 61). Indeed, throughout this book we have discussed numerous examples to illustrate how poor public relations have been the downfall of many organizations that have underestimated the power of the media in shaping public opinion of how a crisis is being handled. One of the best examples of this was the collapse of Gerald Ratner's jewelry empire in the United Kingdom in 1991, which was caused by his throw-away line during a speech to the Institute of Directors in London. Ratner bragged that he could sell a sherry decanter for 15 pounds even if it was "total crap." The next day, the press exposed his mocking insincerity toward the customers that had made Ratner his fortune, and overnight, they abandoned his shops (Weyer 1994).

The media

The construction industry is particularly vulnerable to poor media coverage because of its very negative public image. This was demonstrated in a 1997 national United Kingdom opinion poll, in which only the oil industry was viewed less favorably by teenagers (Building 1997). Furthermore, there is increasing scrutiny of the industry as a result of the ever-greater appreciation of its impact on the built and natural environment (Moodley and Preece 1996). This, coupled with growing sympathies with the environmental movement amongst the general population, has resulted in increasing numbers of confrontations with the public, particularly on road and housing projects. Notably, in many of these increasingly common and public confrontations, the media has portrayed construction companies in a heavy-handed and unsympathetic light and there is little doubt that the future viability of many projects will have been affected by this coverage. In this sense, media relations is an area of traditional neglect to which companies operating in the construction industry must turn their attention. Construction managers cannot rely upon the media to put their case and a continued reluctance to communicate with the media will almost certainly lead to negative reporting of the industry's activities. In contrast, open relationships with the press and more sensitivity to environmental issues will enable managers to better shape the public's attitudes and thereby obtain a more balanced presentation of the facts from the media during a crisis.

One way of ensuring open communication with the press during a crisis is to establish a 24-hour-a-day press office, which has the responsibility of providing factual and up-to-date information to the media and to employees. If managed well, such an office should be able to turn media inquiries into opportunities rather than problems by initiating, rather than reacting to, press, radio, and TV coverage. Public relations are best handled by one trained person who is named as an official spokesperson and who has skills in dealing with the media. TV interviews with untrained staff who appear uncaring, flustered, and unsure of the facts are damaging to the public's perception of competence whereas, a trained person with experience of such events can portray a positive image. The importance of identifying such a person was illustrated during the aftermath of the TWA flight 800 crash when the rush of distraught families, an eager press, and an interested public were left to the management of one chief ticket agent who, through no fault of his own, released inaccurate information which fuelled uncertainty, anxiety, and false speculation about the handling of the affair (Bobo 1997). This was a primary reason TWA was widely criticized afterward by public relations counsellors, crash victims' families, and the media for having an uncaring attitude.

Post-crisis management

After a crisis, a disaster committee should organize follow-up meetings so lessons can be learned and fed into subsequent crisis management efforts. Everyone affected by a crisis must be involved in this process. In addition to managing the learning process, the disaster committee should also turn its attention to the recovery. This can be a lengthy and sensitive process that is likely to be influenced by how well a crisis was managed. For example, it may involve delicate challenges such as conducting investigations into causes, mending damaged relationships, reorganizing the project program, settling on-going disputes and re-assessing project requirements. At the same time, attention must be given to the long-term consequences of a crisis such as rectifying damage to the environment, or dealing with government or legal investigations. Clearly, the less effectively a crisis is managed, the more arduous is the recovery process.

FORMULATING AN EMERGENCY PLAN FOR A CONSTRUCTION PROJECT

Formulating an effective emergency plan takes time and depends on advance planning and the involvement of those who will be affected within and outside an organization. This is often difficult in construction projects due to time and cost pressures, yet perhaps the greatest barrier to emergency planning is the transitionary nature of construction project teams. This requires crisis management plans to reflect the continuous changes in project personnel over time, which has a knock-on effect to training since every change in plan needs to be clearly communicated to those affected. This is best achieved through regular emergency management workshops for everyone and through induction meetings for every new person who enters a project, at whatever level. The complexities and, ultimately, time and costs involved in doing this are a major barrier to the development of emergency plans on construction projects. Despite these practical problems, the importance of planning cannot be overstated. Without a plan, a crisis is likely to stimulate an uncontrollable period of reactionary chaos, which would waste time, resources and energies and inevitably, lead to the deepening of a crisis.

CONCLUSION

While emergency planning can alleviate many of the potential problems that can arise during the crisis management process, it is appropriate to finish this chapter with a warning that managers should not over-rely on their crisis management plans. First, they require continuous maintenance and efforts to ensure that people remain familiar with them. Furthermore, emergency plans cannot cover all eventualities and can be misinterpreted, ignored, or deliberately distorted during a crisis. Ultimately, no matter how well-designed and heavily resourced emergency plans are, *people* are both the weakness and strength of an organization during a crisis. While some people will act selfishly, others will exhibit courage, ingenuity and commitment beyond the call of duty. There is no substitute for good people and responsive management during a crisis.

Chapter 5

Crisis Managers as Social Architects

A construction crisis stimulates a network of communications between internal and external stakeholders. This chapter explores these networks in more detail, arguing that the best crisis managers are social architects, designing, creating, and maintaining appropriate social relationships among project participants.

THE CONCEPT OF SOCIAL ARCHITECTURE

Bennis (1996) introduced the concept of social architecture to encapsulate the contemporary idea that organizations are self-organizing social systems comprising a multitude of interdependent, culturally diverse people with varied and changing interests. Cultural diversity, in an occupational and geographical sense, ensures that these people attribute unique meanings and interpretations to their world and respond differently and unpredictably to managerial actions (Tsoukas 1995). From this perspective, management is no longer seen as a mechanical, regulatory activity designed to control some objective and static scene, but as a social activity that involves coordinating purposeful individuals who are embedded in complex and constantly changing social networks. These networks tie discrepant parts of an organization into a coherent whole by providing channels through which information flows and new ideas are spread. In this sense, they are the central mechanism through which mangers must work to achieve integration, coordination, and cooperation. In practical terms, this view of organizations implies that managers should focus on relationships rather than on individuals and that they need to understand the dynamics and structures of these social relationships and how they influence an organization's ability to achieve its goals.

SOCIAL STRUCTURE AND CRISIS MANAGEMENT EFFICIENCY

The first evidence to support a link between the structure of people's relationships and crisis management efficiency was provided by the pioneering work of Bavelas (1950), Leavitt (1951) and Shaw (1954). They assembled small groups of people, requiring them to solve a simple problem that required the pooling of information. The members of each group were physically separated and only permitted to use predetermined communication channels, as illustrated in figure 5-1.

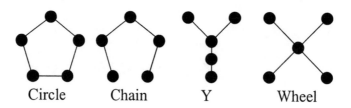

Circle   Chain   Y   Wheel

Figure 5-1 Leavitt's experimental patterns.

Through observations, interviews, and analysis of each group's communications, the researchers found that there were significant differences among the communication patterns in terms of their problem-solving efficiency. When problems were simple, organizational efficiency developed most rapidly in the chain followed by the Y, the wheel, and finally the circle. In terms of leadership, the circle was seen as active, leaderless, unorganized and erratic, yet enjoyed by its members. In contrast, the wheel was relatively inactive, well-organized, less erratic, and unsatisfying to its membership. The researchers used the concept of "centrality" to explain the results. Centrality is defined as "the degree to which information flow is centered around one or a few individuals in a communication network." Low-centrality organizations are characterized by no leader, high activity, slow problem-solving, but high satisfaction. However, the problem-solving efficiency of each network depended on the nature of the problem faced by a group. When the problem complexity increased, the relative efficiency of various patterns completely reversed; the circle became the fastest and the wheel the slowest. The central person in the wheel became overloaded with information and peripheral people were less willing to merely accept the solution offered by the central person.

Although Leavitt and Shaw conducted their experiments in an artificial environment where all the complications of real-life were stripped away, their findings suggest that crisis management efficiency would be influenced by management's ability to control the structure of people's communication patterns. Therefore, it is worth considering further the various dimensions of communication structure that are likely to be important in a crisis management setting.

Formal and informal communication structures

At the most basic level, communication structure can be categorized as formal or informal. Formal communications during a construction crisis are determined primarily by employment contracts that bind project participants together. These contracts set down prescriptive procedures designed to dictate the patterns of interaction of organizational members, thereby imposing an element of predictability in a situation where managerial control is naturally threatened. Such procedures are also designed to ensure a rapid response to crises by keeping information channels free from irrelevant information and by defining the boundaries of acceptable behavior. In this sense, it appears that construction contracts have a significant influence on the actions of project members during a crisis. However, Sagan (1991) argues that the influence of contractual rules and procedures diminishes during a crisis. His research indicates that standardized procedures are most helpful when the nature of the task is simple whereas, when task complexity and time-pressures increase, they become restrictive and counter-productive by removing people's autonomy. Furthermore, Bax et al (1998) indicate that during a crisis, a gap can open between what workers consider to be a legitimate and effective way of dealing with a situation and what formal procedures prescribe. This is particularly likely if workers have not helped formulate those rules. As Loosemore and Hughes (1998) found, the tendency in such situations would be for formal procedures to be bypassed and for people's patterns of communication to revolve around common interest groups and friendships.

The dangers of contracts as a means of control

Thus, the network of communications that emerges in response to a construction crisis has a significant informal element that could be dangerous if managers relied too much on contracts as a means of control. Contracts, however well drafted, are no substitute for good management. Rather, managers facing a construction crisis must ensure that formal procedures complement informal systems and recognize that a successful outcome may depend on their willingness to release people from potentially restrictive rules and procedures. This demands some courage, since there is a natural tendency to turn to contracts during a crisis in order to re-impose control. It also demands a certain degree of confidence and trust in fellow project members, which may not exist due to the organizational practices and historical divisions discussed in Chapter One. Indeed, under the traditional confrontational environment that pervades many construction projects, a sudden movement to a more flexible stance may be disastrous because it would be alien to the expectations of individuals who have developed behavior patterns ideally suited to an untrusting environment. The likely result is that people will see such a cultural change as an opportunity for exploitation rather than for increased cohesion. This is precisely why a number of projects using the more flexible Engineering and Construction Contract (1995) produced by the Institution of Civil Engineers in the United Kingdom have not been as successful as hoped.

Thus, it would seem that the voluminous, rigid, and prescriptive nature of traditional construction contracts is a necessary response to the untrusting and confrontational environment that pervades many construction projects. Ironically, such contracts perpetuate this environment, indicating that the construction industry is caught in a downward spiral of mistrust and control that is increasingly difficult to stop. This problem led Loosemore and Hughes (1998) to propose "the emergency option" as an intermediate step in making traditional contracts more flexible with a minimum risk of exploitation. In essence, the emergency option is a separate clause that can be legally incorporated into traditional construction contracts, enabling parties to mutually agree to opt-out of normal contractual procedures for a specified period in order to cope with a crisis. If one party abuses this trust, then there is a facility for any party to unilaterally return the project to a formal mode.

Groups, factions, and cliques

People are social animals, spending much of their time in formal and informal groups that have their own unique norms (expected standards of behavior), agendas, orientations, and cultures. Groups perform many important functions, particularly during a crisis, such as enabling people to solve tasks of much greater magnitude and complexity than they would be able to tackle alone. Groups also provide for people's social, identity, and belonging needs and can represent an important source of power in negotiations. An awareness of group formation is an important aspect of crisis management. However, Tichy et al (1979) argue that the roles people play within and between factions are also important to managers. For example, a "liaison" is an individual who is not a member of a faction but who links two or more factions, and a "bridge" is an individual who is a member of multiple factions. Liaisons and bridges perform important linkage functions in organizations and their removal destroys its connected unity. Their identification by a crisis manager is particularly important because they can be used to prevent communication breakdowns and to mediate among groups of opposing interests before misunderstanding and conflict emerge. A third inter-faction role that may be valuable during a crisis is the "isolate," who is not affiliated to any one faction in particular. Isolates may be crucial sources of independence that crisis managers can use to their advantage, but the danger is that in the heat of a crisis, powerful factions can dominate communications and divert attention away from them. Therefore, managers may need to make special efforts to integrate isolates into the crisis management process. Indeed, by increasing the power base of isolates or by focusing on them, crisis managers can minimize the potential for conflict arising from the competing interests of different groups.

The dangers of groups

While groups can perform many useful functions during a crisis, they can also be at their most dangerous. For example, Hornstein (1986) has expressed alarm at the in-roads being made by the social ethic that espouses groups, not individuals, as the prime source of creativity and that proclaims group membership, an experience akin to being in a family, as the ultimate need of an individual. In Hornstein's view, groups have a capacity for producing a special, perniciously subtle tyranny that can damage communications, slow down decisionmaking, produce compromise decisions, and suppress creativity and innovation. Marsh et al (1978) also noted the powerful influence a group can exert over its members. Their research into football hooliganism concluded that group norms can cause people to blindly exhibit almost tribal behavior that their personality and wider society would normally suppress. Janis (1988) refers to the tendency of groups to emphasize the importance of consensus and agreement as "groupthink." The potency of this effect depends on the attractiveness of a group to its members and the extent to which conformance enhances a group's power, aids its survival, benefits its members, simplifies its processes, and expresses its central identity and values. The potential danger of groupthink was vividly illustrated in the Challenger Space Shuttle disaster when, on the evening before the launch, engineers, with full knowledge that the expected temperature at the time of launch would be below safety levels, were pressured by peers to sanction the launch. It would seem that during a crisis, when effective communication, open-mindedness, creativity, and flexibility are most valuable, groups could be at their most dangerous.

Structural equivalence

Another aspect of social structure that is relevant during a crisis is peoples' structural equivalence. Two people in an organization are structurally equivalent if they have exactly the same pattern of contacts. At first it was thought that such people play the same role in an organization and were inter-changeable with one another. However, the concept of structural equivalence only considers similarities between people's "patterns" of interaction and ignores the nature of people within them. That is, people can be structurally equivalent by having the same patterns of connections with different people. Scott (1991) argues that this does not mean that they are playing the same social role and advocates the alternative concept of "regular equivalence."

Regular equivalence

Two people are regularly equivalent if they are connected to the same people in the same way. The level of regular equivalence in an organization is important to crisis managers because it is reasonable to assume that communication would be better and the crisis management process more efficient within highly equivalent groups. The members of such groups would have common neighbors, shorter communication routes, and higher levels of communication.

While highly equivalent groups seem desirable, they could be particularly susceptible to groupthink. Variety in social relationships is clearly important in generating the creativity needed to resolve a crisis. Therefore, it would seem that crisis managers have a "tight-rope" to walk in balancing the level of regular equivalence and independence they encourage.

Centrality

Organizational centrality is the degree to which information flows are centered on one or a few people. The concept of centrality is important to crisis managers because substantial evidence suggests it may be closely related to crisis management efficiency. Leavitt (1951) and Shaw (1954) found that the influence of centrality depended on the nature of the problem an organization faced—more complex, non-routine problems such as crises demanded less centralized structures to alleviate the potential for information overload. However, Mintzberg (1976) found that during a crisis, people tend to tighten control, the consequence of which is dysfunctional behavior. It is a paradox of crisis management that if inappropriately applied, centrally imposed order can lead to disorder where disorder would eventually lead to order.

Thus, the concept of centrality is important to crisis managers who must seek to control the degree to which their organization is centralized around particular individuals. This is best achieved with an understanding of the different types of centrality that exist. Freeman (1979) is credited with clarifying the literature in this area. He differentiated between the concepts of *degree centrality*, *closeness centrality* and *betweenness centrality*, each having sharply different implications for the management of crises.

Degree centrality

Degree centrality refers to people's roles as senders or receivers of information within an organization. A person with a relatively high sending role is a prominent *source* of information to others, is in the "thick of things," is a highly active member of the network and is a focal point for instruction and leadership. Such people sustain an organization by providing the information that is its "life-blood," and in this sense, an organization is highly dependent on them. This places them in a powerful position, particularly during a crisis when people's appetite for information naturally increases. In contrast, people with a relatively high receiving role are prominent *sinks* of information and are likely to play an important information storage or synthesizing role within an organization. They are especially important during a crisis because of the great volume and variety of information that is generated and that needs to be synthesized, converted, or condensed into a manageable, meaningful, and consistent format

Betweenness centrality

Betweenness centrality measures the extent to which a person lies between others in an organization and reflects the degree to which they play an information "gate keeping" role. People with high betweenness centrality are important because they have the capacity to manipulate or filter information flowing between people. In this sense, they have a great deal of power, acting as the valves within a network and occupying a critical position in maintaining free and open communication. Essentially, these people act as the glue that holds an organization's parts together, and weaknesses at these critical points can lead to disintegration.

Thus, an organization's betweenness centrality is an important measure of its vulnerability to people's integrity in not manipulating information to satisfy their own ends and to people's ability to manage the information passing through their hands. These vulnerabilities are naturally exacerbated during a crisis because high stakes cause people to pursue their interests with more tenacity and because organizations can become flooded with information at the same time as high stress levels reduce people's information handling capacities.

Closeness centrality

People have a high closeness centrality if they are positioned at short "distances" (measured by the number of intermediaries) from every other person in an organization. The closeness of a person to all others is a reflection of his or her independence since high closeness makes it difficult to act alone, without others knowing. Conversely, people with a high closeness centrality have the capacity to directly monitor and control others and to communicate their ideas more rapidly to a wide audience with minimal distortion. During the pressures of a crisis they represent a very important channel of communication for a manager.

CRISIS BEHAVIOR

In the midst of a crisis, managers must control people's patterns of communication because the patterns determine the ease with which information flows within an organization and thereby, the level of uncertainty, misunderstanding, and ultimately, conflict that arises. In this section, we explore the unfortunate tendency for crises to cause people to behave in ways that make this difficult. This behavior is largely a consequence of two forces: people's difficulties coping with the pressures and stresses of a crisis and their difficulties coping with the significant levels of change induced by a crisis.

Psychological pressure and stress

By definition, crises are potentially serious events that require inventive solutions under extreme pressures. This ensures that those affected feel a certain degree of tension and anxiety—conditions that induce both positive and negative behavior. For example, while some argue that such feelings produce a determination that is important to the efficient resolution of problems, others point to them increasing suspicion and reducing communication. The explanation for this contradiction lies in the distinction between pressure and stress. Robertson and Cooper (1983) argued that *pressure* is a force acting on an individual to perform in a particular way or to achieve a particular end result. It can be a source of some discomfort and anxiety but at the same time it can be exciting, challenging, and growth- producing. On the other hand *stress* has only negative outcomes for an individual because it arises from an inability to cope, which produces defensive and maladaptive behavior.

The impact of stress on crisis management outcomes

According to George (1991), stress is a "generic problem that poses severe threats to crisis management" (p. 559). The stress associated with a crisis arises from the dramatic challenge to previously held views, from the dislocation to social relations and from the physical challenges posed. Stress can also come from the psychological shock of a crisis, which can manifest itself immediately or in the form of post-trauma. An example of a crisis that could cause this type of shock is a workplace death. A co-worker's death on the job could devastate an organization's employees. However, the impact of stress is never uniform in its effect and those who were physically or emotionally close to the individual would be more likely to experience difficulties. Furthermore, during a crisis like this, only higher levels of management have the authority to deal with the enormity of decisions required. This means that the burden of pressure would fall upon their shoulders and it is the people occupying these positions of responsibility who are also in particular danger of suffering stress.

The behavior resulting from stress is precisely what is not needed during a crisis. Hermann (1963) pointed to a loss of attention to problem-solving, increased decisionmaking errors, greater rigidity in exploring alternative courses of action, panic, anxiety, and withdrawal. Stewart (1983) found that individual reactions to stress include agitation, reduced attention span, absenteeism, sickness, aggressive behavior, impulsive behavior, depression, lower tolerance of risk, and lower tolerance of other's opinions. T'Hart (1993) reported that stress also amplifies personal insecurities and feelings of vulnerability and may decrease the self-confidence and self-esteem of those affected. Finally, in extreme circumstances, stress can seriously impair psychological well-being and might even activate latent psychological vulnerabilities or borderline psychopathological tendencies (George 1991).

In this sense, managers must appreciate that the costs of a crisis are not all physical and that its psychological impact can paralyze an organization by traumatizing its employees. Indeed, Kutner (1996) points out that stress-related disabilities now account for 14 percent of worker's compensation claims and are twice as costly as the average physical injuries claim. Clearly, the location of "stress-points" within an organization is something to which crisis managers should give serious attention. To alleviate potentially problematic behavior it is critical that sufficient consideration is given to the personalities and capabilities of people occupying these points and that adequate support is provided for them. This process should be a continuous one that should extend beyond the solution of a crisis because the effects of stress are often delayed in their impact and the location of stress points is constantly changing.

Coping with change

Crises inevitably induce a significant amount of social, monetary, and physical change that most people find unsettling because it represents an abandonment of past efforts and a threat to the status quo. This often produces resistance to change that can take many forms, ranging along a continuum from passive disagreement to positive hostility. The level of resistance is likely to depend on the extent of change required, the extent to which people's interests are damaged, the power of those whose interests are damaged, and the manner in which change is introduced. Ansoff (1979) argues that while change may eventually result in resistance, its emergence might be delayed by the inherent seriousness of a crisis. For example, if a crisis is serious enough to threaten the very existence of an organization, the initial response may be to put differences of interest aside, to tackle the crisis and thereby ensure the organization's survival, which is in everyone's interests.

The growth of a conflict

While resistance may be the natural response to change, the conflicts of interests that exist within most organizations complicate people's behaviors. Changes that are a threat to one party will be an opportunity to another. While managers can empower these potential allies to champion the changes that need to be implemented, they also have to deal with the tensions created between potential beneficiaries and losers because it is within such tensions that conflict is born.

Conflict is a progressive phenomenon that gathers momentum as it escalates through the phases of simple disagreement, contention, dispute, limited warfare and all-out warfare, where parties are trying to destroy each other at all costs (Philips 1988). The natural tendency for crises to generate conflict has caused Snyder (1972) to describe a crisis as a transition zone between peace and war where the speed of transition depends on factors such as the gap between opposing parties, past and current relationships, attitudes toward compromise, and the way bargaining is managed. Since bargaining is the initial and informal means by which people attempt to resolve their differences, an understanding of bargaining processes should represent the foundation of a crisis management strategy. A better command of this process should reduce the possibility of escalation and the associated movement toward more formal, costly, public, and time-consuming methods of resolution such as arbitration and litigation (CME 1997). As Pinnell (1999) argues, conflict management skills are like an insurance policy against catastrophic loss, regardless of that project's complexity and size. This was well illustrated in two recent construction disputes that went wrong. One occurred during a simple $600,000 sewer contract in Arizona and resulted in the plaintiff being awarded $300,000 in damages and both parties having to pay $257,000 in costs. In another more complex case, the attorney's fees alone exceeded $5 million.

The bargaining process

The term "bargaining" implies a difference in interests, objectives, and expectations and is concerned with reaching accommodations between them. More precisely, bargaining is "a process whereby parties negotiate over the distribution of scarce resources, money, status or power" (Morley 1981, p. 113). In essence, the process involves a struggle between adversaries who attempt to move, step-by-step toward an agreement over resource redistributions that is in their own favor. In this struggle, which is like a game of chess or poker, people employ a range of tactics that, in a construction project, may be motivated by a complex web of interpersonal and inter-organizational forces. This is because negotiators belong to and represent the interests of distinct profit-making organizations that ensures that their attitudes and behavior is determined not only by their own values, but by those of their employers and of any informal interest-groups or temporary coalitions to which they are affiliated. Indeed, some employers may impose real restrictions on a negotiator's decisionmaking authority and autonomy at the bargaining table, making it difficult to resolve issues quickly.

Bargaining tactics

The essence of the bargaining process is the "tactics" or "moves" negotiators use to influence each other. Rogers (1991) argues that the tactics adopted by an individual depends on their "bargaining code"—their set of beliefs about an opponent that influence the way they interpret and respond to their messages. According to Rogers, the beliefs that are important in a bargaining setting are those relating to *an opponent's objectives,* to *dispute dynamics* (i.e., the manner in which war might erupt) and to *the optimal mixture of coercion, accommodation, and persuasion in a bargaining strategy.* On this basis, Rogers grouped bargaining codes into four broad categories: types A, B, C, and D. The characteristics are depicted in Table 5-1 and lead to the employment of certain tactics.

Tactical miscalculations and accidental escalations

Table 5-1 illustrates that during bargaining, negotiators make tactical choices guided by various beliefs. Unfortunately, under the pressures of a crisis, negotiators are often forced to make decisions with incomplete information and consequently there is a chance of tactical miscalculations that can precipitate an unintentional escalation of a dispute. Many vivid illustrations of this danger can be found in the area of international relations. For example, America's war with Japan in 1941 was precipitated by a U.S. oil embargo and inflexible demands for Japan to absolve claims to sovereignty in Asia. This forced the Japanese into a corner and gave them no alternative other than to initiate war by making a pre-emptive strike in Pearl Harbor. Other, more recent examples of conflicts precipitated by tactical miscalculations are the wars between South Korea and North Korea in the 1950s, the Falklands war in the 1980s, and the Gulf War in the 1990s. In each case, the aggressor served a *fait accompli* upon its opponent, failing to predict the nature and intensity of their response. The South Korea/North Korea war illustrates the difficulties in predicting an opponent's response, even with good intelligence gathering, since the U.S. unexpectedly reversed its earlier policy of non-intervention and quickly came to the assistance of South Korea.

As Dixon (1988) points out in his psychological analysis of military incompetence, the likelihood of tactical miscalculations becomes an increasing danger as a dispute escalates, which means that they tend to gather momentum, once initiated. This is because people's perceptions of each other become less rational and more guided by emotions. The most frightening illustration of the fragility of conflicts, when they escalate, occurred on October 25, 1962 when nuclear armed bombers sat on runways around America ready for war with the Soviet Union. Pilots had been told that there

Table 5-1 Bargaining codes and tactics (Adapted from Rogers 1991).

CODE	BELIEFS
A	**Adversary**: Seen as aggressive. **Dispute dynamics**: Only intentional war is possible. Little consideration needs to be given to the response of an adversary and its escalating impact. Any escalation is easily controllable. **Tactics**: Open use of aggression, *fait accompli,* or strong coercive action is the best way to resolve a dispute. Success in negotiations is best measured in military rather than diplomatic terms.
B	**Adversary**: Likely to employ offensive, rather than defensive tactics. **Dispute dynamics**: Control of a dispute is possible to a point where unintended escalation is possible. It is possible to understand the dynamics of escalation and thereby avoid the point where control is lost. Probabilities of escalation can be assigned to various tactics and strategies. **Tactics**: Incremental small-step escalations will be seen as timid and a sign of weakness and are likely to lead to an escalation. Failure to show resolve is the most common cause of war. **TYPE B-1** **Adversary**: Willing and able to take advantage of any signs of weakness. **Dispute dynamics**: Escalation is assumed to come from failure to communicate a determination to protect one's vital interests at any cost. **Tactics**: Coercive diplomacy (i.e. verbal threats of extreme actions and all-out war) and bluffing are the best means of dispute resolution. It is dangerous not to brandish the ultimate weapon. **TYPE B-11** **Adversary**: Seen as unpredictable. **Dispute dynamics**: Bluffing and threats are dangerous since they may inadvertently induce a counteractive aggressive response. **Tactics**: It is better to use limited force to avoid an all-out war rather than to use threats of all-out war.
C	**Adversary**: It is difficult to determine whether adversary is offensive or defensive. **Dispute dynamics**: Two images of escalation - failure to show resolve and spiralling responses to perceived provocations. Many unpredictable paths to escalation, difficult to avoid slippery slopes, brink cannot be recognised in advance. **Tactics**: Tactics must be cautious and context-driven rather than automatic. Only partial control of dispute is possible and threats or use of power are dangerous. Tread carefully, limited escalations and compromises preferred. Carrot and stick approach is best means of manipulating an adversary.
D	**Adversary**: Assume adversary operates in a defensive mode. **Dispute dynamics**: Control of disputes is very problematic if not impossible with even a modest emphasis upon coercion. Highly cautious approach in fear of triggering an uncontrollable escalation. **Tactics**: Accommodation and compromise is best means of resolution. Entire effort should be aimed at avoiding bargaining situations.

would be no practice drills during this tense crisis and when a sentry in one military base spotted someone climbing a fence, he suspected a soviet saboteur and sounded an alarm. The intruder was, in fact, a brown bear but in one base, the wrong alarm was sounded and pilots were sent running to their aircraft, fully believing that a nuclear war was beginning. Only when the base commander realized the mistake and drove onto the runway to prevent them taking off was the mission to bomb the Soviet Union aborted.

In addition to the problem of irrationality, which can fuel a dispute, there is the problem of increasing inflexibility. This occurs as people become progressively embedded in their own position as increasing investments of resources reduce their willingness to compromise. Under these conditions, winning at all costs becomes increasingly important and people become caught up in a self-perpetuating spiral of conflict that becomes increasingly difficult to break and that sucks in unjustifiable quantities of resources. This was vividly illustrated in the Vietnam War where military commanders were responsible for executing policies that cost the United States $300 billion. During this war, which achieved nothing in strategic terms, almost 2 million people died and the U.S. released 13 million tons of high explosives (more than six times the weight of bombs dropped in the whole of World War II).

The dynamics of conflict – the use of inducements not threats

In conflict resolution, parties attempt, by negotiation or other means, to force an adversary to come to the table, to make concessions, and to accept an agreement that meets their interests and needs. Third parties also influence the process by backing one party, mediating between them, or by manoeuvring to protect their own interests. This process of trying to influence an opponent normally relies on a series of threats and inducements. For example, in construction projects, a threat could be to withhold payments, an inducement to trade-off a settlement against other outstanding disputes. In most construction projects, the use of threats far outweighs the use of inducements, yet inducements are particularly effective if they meet the needs of an opponent and encourage reciprocations that can transform the landscape of a bargaining setting. An example of this occurred in 1977 when Egyptian president Anwar Sadat visited Jerusalem and made an unexpected concession in accepting Israel's position in the anticipation that they would reciprocate. This initiative was designed to transform a seemingly unresolvable conflict into a new constructive relationship that would enable the stalemate to be resolved by peaceful means. As Kelman (1997) points out, while an emphasis on positive inducements rather than negative threats is more risky in the short-term, it has the potential to be far more positive in the long-term. However, the effective use of positive inducements requires more than just offering an opponent the rewards and promises that are most readily available. Rather, it depends on the use of inducements that address an opponent's fundamental needs and fears and on their willingness to reciprocate.

Unfortunately, the high levels of understanding, altruism, and trust needed to initiate this process are lost rapidly as a conflict escalates and third party intervention is often required. A key aspect of this person's role is to facilitate mutual reassurance, which can be brought about by acknowledgments, symbolic gestures, and confidence-building measures that demonstrate mutual sensitivity. Such gestures need not cost any party a great deal. For example, it may be necessary for one party to simply acknowledge past mistakes or agree to meet. Such gestures have a powerful psychological impact in opening the way to negotiations, even though they may not be immediately transferable into concrete actions. This is because most conflicts are marked by a history of accusations and denials of another's experiences, authenticity and legitimacy. For example, at the beginning of Sadat's visit to Israel, Sadat offered a symbolic gesture that had a disproportionately large impact upon the Israeli negotiators: he offered to shake their hands. Previous officials' refusal to do so had come to symbolize Arab denial of Israel's legitimacy and the very humanity of its people. Thus, positive incentives, if they are genuine and well-thought through, have an advantage over threats as a means of conflict resolution because they provide the basis for the building of new, more positive relationships, which can become an incentive in their own right. Once this process has been initiated, parties will be inclined to live up to each other's expectations in order to maintain and extend the relationship. If the relationship blossoms, then adversaries will be able to approach their conflict as a shared dilemma and are more likely to reach an effective resolution.

Conflict as a positive force

The traditional view of conflict in the construction industry is as a disruptive force that should be avoided and eliminated at all costs. However, in Chapter One we saw that conflict is inevitable in organizations and that it can have positive implications if managed effectively. A well-managed conflict can force a more thorough investigation of a wider range of crisis solutions and can act as a useful release-valve for accumulating tensions that would otherwise remain concealed. In this sense, the potential for conflict during a crisis must not be seen as entirely destructive. However, Loosemore et al (1999) found that the skills and attitudes to manage conflict constructively do not yet exist within the construction industry. In this sense, the encouragement of conflict would seem premature. However, this is no justification for the current trend toward reducing construction conflict at all costs. This would incur significant opportunity-costs for the construction industry by reducing the possibility of benefits arising from the effective management of construction conflicts. In the long-term, a more intelligent and beneficial strategy would be to change people's attitudes to provide the foundations for constructive conflict management. The problem for the construction industry is not necessarily in the existence of conflict but in the way it is managed.

PHASES OF BEHAVIOR DURING A CRISIS

The previous discussion highlighted the different types of behavior that might be expected during a crisis. However, it did not identify when, during the life of a crisis, certain types of behavior might evolve. For example, Cisin and Clarke (1962) proposed a three-stage model of *impact, reaction,* and *reconstruction*. During the period of *impact* people can do very little to deal with the crisis. Behavior is primarily aimed toward survival and unintelligent behavior can significantly contribute to a magnification of the possible losses to the individual and to the community as a whole. The *reaction* period is one of maximum disruption characterized by immense communication difficulties and deviant behavior. It is a phase of damage assessment, high anxiety, and confusion in which people behave irrationally and inappropriately. The main managerial problem is one of coordinating different individuals who tend to act individually and on their own definitions of the problem. Finally, the period of *reconstruction* is the beginning of the return to normality where the damage done is put right.

Fink et al (1971) produced a more detailed model that also showed patterns of behavior evolving in a predictable order. They argued that initially, crises have a disorienting effect that induces a sense of panic, disorganization, and chaos. Once this period of *shock* has subsided, people's minds turn to self-preservation and a period of *defensive retreat* where people seek to protect their own interests and to maintain the status quo. Furthermore, interpersonal relations become inwardly orientated as people turn to the protection of their interest groups. Unfortunately, this behavior deepens existing divisions and afraid of losing control, leaders tend to centralize decisionmaking power and information flow. However, this only divides an organization into factions and ritualizes communication to the extent that information is exchanged without any useful exchange of meaning. As it becomes increasingly apparent that the crisis cannot be resolved in this way, the process moves onto a phase of *acknowledgment*, self-examination, and interpersonal confrontation. During this stage, psychological stress and tension are high and there is the danger of communications degenerating into accusation and blame that can rapidly lead the organization back into the *defensive retreat* phase. However, eventually, in a need to solve the problem, individuals will resolve their differences and move to a more constructive orientation of *adaptation and change*. In extreme circumstances, this may require the involvement of a third party and the intention is to search for better ways of communicating which lead to genuine understanding and a meaningful sharing of information. As this occurs, leadership becomes more relaxed and there is a greater emphasis upon collective decisionmaking. Furthermore, people work more interdependently and let go of the dysfunctional behaviour that characterised the early phases of the crisis. Eventually, inter-group relations become coordinated and the organization once again begins to resemble a stable state.

Finally, Sipika and Smith (1993) propose a three-phase model. The first phase is *the crisis of management* where the organization's culture serves to incubate the crisis until it becomes unsustainable and a trigger event propels the organization into the *operational phase* that is typified by convergence and the presence of high energy levels. Confusion reigns as the level of complexity in communication increases in an attempt to cope with the crisis. Eventually, the organization moves into the *legitimization phase* where a recovery strategy is developed.

The value of these models is not in the detailed behavior they describe but in the behavioural dynamics they depict. This illustrates that crisis management strategies need to be as dynamic as the behaviour they seek to control.

CONCLUSION

This chapter has discussed the ways in which people tend to behave during a crisis. The various principles that have been introduced will be used in the following chapters to analyze the management of four real-life construction crises. The data that form the basis of each case study was collected from diaries that were completed by project members before, during, and after each crisis, from observations of project meetings, from documentary inspection, and from retrospective interviews with project participants. They are both amusing and shocking and provide revealing insights into life on a typical construction project.

Before progressing, it is important to point out that the crises that form the basis of the following case studies occurred during one of the worst construction recessions in living memory. This undoubtedly affected people's behavior. Furthermore, the case studies do not describe the aftermath of construction disasters that made the headlines. The purpose of this book is to help managers avoid such disasters through a better understanding of crisis management skills. Instead, the case studies describe the development of four construction crises that posed a serious threat to the viability of their host projects. They are typical of the crises experienced on many construction projects at some time during their life, and although they did not make the headlines, there are responsible for the majority of the cost and time overruns in the construction industry.

Chapter 6

Case Study One

INTRODUCTION

The next four chapters will refer to professions that may be unfamiliar to some people. For example, the quantity surveyor (QS) creates and controls budgets at various project stages. On most projects a consultant QS represents the client's interests and a counterpart works for the main contractor. Traditionally, in the pre-contract phases of a project, the consultant QS produces a "bill of quantities" that itemizes operations involved in the construction of a building. This document is priced by tendering contractors and used as a basis for comparing their tenders. Once construction has commenced, it is then used to value construction work for payment purposes.

The clerk-of-works is another role that may be unfamiliar. This person has the responsibility to monitor site activities during construction and to report back to the architect on a regular basis. In essence, the clerk-of-works is the architect's eyes and ears on site, although with limited contractual power.

THE PROJECT

The project that represents the basis of this case study was a major extension to an existing leisure-centre on a difficult, constrained site that bordered a major road. The main contractor and all consultants had been employed on the basis of the lowest bid on a competitive tender, and coincidentally, had worked together on a recently completed project. During this project, relationships had become strained and the contractor had gained a reputation among the consultants for being "claims-conscious."

THE CRISIS

A creeping crisis began when, during basement excavations, the main contractor encountered an unexpected problem. In the contractor's opinion, permanent earthwork support was needed to construct the basement wall in this area adjacent to the road, but the bill of quantities had only enabled them to price for temporary earthwork support. The contractor contended that the QS should have made the provision in the bill of quantities for permanent earthwork support and that consequently, there was an entitlement to extra payment. The consultants disagreed. Polarized positions combined with the contractor's cessation of work on critical path activities led to an acrimonious dispute that lasted for approximately 10 months and eventually resulted in serious delays, costs, and the replacement of the contractor's entire site team.

AN ACCOUNT OF THE CRISIS MANAGEMENT PROCESS

A chronological account of the crisis follows. It is divided into separate periods of activity to highlight the dynamics of the crisis management process.

The contractor states an intention to claim extra payment

During a site meeting, the contractor's QS stated his intention to claim for extra payment to cover permanent earthwork support. He warned that work would stop in the affected area until the claim was sanctioned.

The contractor's justification for this tactic rested on his distrust of the architect. This distrust had also delayed notification of the bill of quantities discrepancy that had been apparent to the contractor for some time, but was withheld until the last minute to pressure the consultants into making a quick decision. The contractor's distrust of the consultants was mutual. The client's QS was suspicious of the contractor's motives in making a claim because of their very low tender, ambitious program; claims-conscious reputation; and a concern previously expressed by the contractor, that the consultants' excavation rates had been under-priced.

The consultants reject the contractor's claim

The contractor's claim for extra payment was formally rejected by the architect who simply stated that alternatives to the contractor's suggested permanent earthwork support system were available that would fit within the scope of the original bill of quantities description. Although the consultants' had not established alternative systems, the initial tactic was to make the "*bald*" statement that one existed. The aim was to send "*a clear message to the contractor that claims would not be tolerated*" and to "*test the contractor's resolve*" in pursuing the claim.

The consultants construct an opposing case

While waiting for the contractor's response, the client's QS and engineer constructed an argument to discredit the contractor's proposed earthwork support system. The client's QS admitted to "*guiding the engineer to look for alternatives because if there was an alternative, however expensive, then it was covered by the bill description.*" The contractor's site manager was aware of this tactic because the engineer, uncomfortable with being coerced into suggesting unreasonable solutions and being sympathetic with the contractor's case, had confided in him, "blowing the whistle" on his fellow consultants. This exposed divisions within the consultant's team and made the contractor's site manager pursue his claim with greater tenacity.

The client's project manager is alerted to the on-going dispute

The client's project manager was alerted to the claim by the contractor but was reassured by the consultants that it had been rejected.

The contractor formalizes the dispute

The contractor's site manager eventually wrote to the architect, complaining about the ongoing dispute, warning of accumulating delays, and threatening further action. This was a significant step in formalising the dispute, yet, the opportunity to discuss the problem in the next site meeting was not taken, the architect merely making a formal statement that "*alternative systems were being investigated by the engineer*". The architect's rationale for not discussing the dispute was that it was not an appropriate forum. The contractor's site manager offered a more cynical explanation that reflected deteriorating relationships: "*[the architect] did not particularly want it minuted and he didn't feel capable or want to talk about it. There was no point me pursuing it with him because if you are going to talk to someone then you have to do it in a fairly positive way.*" The contractor's site manager wrote to the architect again, reiterating his warnings of accumulating delays on site.

The contractor bypasses the architect

The contractor's site manager bypassed the architect by directly telephoning the engineer to discuss alternative earthwork support systems. The engineer suggested ground-freezing but both agreed it was unreasonable to classify this as a temporary earthwork support system as described in the original bill of quantities. Despite this, the contractor's claim was rejected once again in the next site meeting on the basis that ground-freezing was a viable alternative.

A second problem arises

The client became involved for the first time because heavy rain undermined a water main that had been exposed by the excavations and left unattended due to the on-going dispute. This required the diversion of the main and the temporary cutting-off of water supplies to the existing leisure center, which was still in use.

The dispute escalates

The sudden involvement of the contractor's regional surveyor escalated the dispute. This was a desperation tactic by the contractor's site manager who felt that "*relationships had become so poor that the problem could not be resolved at site level.*" The regional surveyor immediately telephoned the architect and client's project manager, threatening litigation. This prompted the consultants to verbally sanction the extra payment.

Payment is refused

The client's QS asked the client's project manager to formally sanction the extra expenditure, but he refused, requesting evidence that all alternative earthwork support systems had been considered. He was under the impression that the claim had been rejected and was surprised that there had been an on-going dispute. He was also annoyed at being excluded from preceding communications and was suspicious that his fellow consultant's had done this deliberately: *"I was informed that the claim had been rejected and as far as I was concerned, nothing had changed. It's a cynical view but I think there had been a lot of work going on behind the scenes to cover it up. We will often contra-charge consultant omissions against their fees."*

The evidence provided by the client's QS was not satisfactory to the client's project manager and he requested a verbal explanation of events. This further delayed the sanctioning of the claim and frustrated the QS who felt that the architect was trying to distance himself from the problem.

Eventually, the client's project manager sanctioned the extra payment, although discussions still continued between the client's QS and engineer about the possibility of alternative earthwork support systems.

A third problem becomes evident

More severe rain caused the unsupported excavations to collapse, making un-viable, the contractor's newly sanctioned earthwork support system. After a period of relatively intense communication among all parties, the contractor proposed another new earthwork support system that cost less than that which was originally proposed. The rates for payment were amicably agreed upon, although the client's QS refused to permit the contractor the benefit of re-negotiating rates at a higher level than they had under-priced in their original tender.

The contractor serves a second claim for loss and expense and an extension of time

The contractor served a second claim for extension of time and loss and expense arising from the delays in sanctioning their first claim. This had been planned for some time, but had not been issued in fear of jeopardizing the original claim.

The architect ignores the claim

The architect's initial response to the contractor's second claim was to ignore it, which prompted the contractor's QS to re-serve it. Again, there was no formal response from the architect, apart from pronouncing its rejection during the next site meeting. There was no opportunity for discussion and no reasons were given. The architect explained that *"in my view the whole problem is resolved, they will go on and on restating their demands and won't get anywhere and so it will go on until completion of the project, to the final account and so on and so forth and ultimately they will go away*." This meeting was made more tense when the client's project manager complained that the contractor's delay in making the claim made the monitoring of project progress difficult: "*Perhaps I am a bit cynical but I do feel that problems are kept quiet so that they can rumble on to give them something concrete.*" This complaint prompted a defensive response from the contractor's QS, who maintained that contractually they were not obliged to notify about potential problems, only actual ones.

The second claim is formally rejected

The architect formally refused the second claim but with no detailed explanation. After two months, the contractor's regional surveyor intervened again. This was prompted by the architect writing to the contractor's directors complaining about the "*unprofessional conduct*" of their project team. An acrimonious exchange occurred between the regional surveyor and architect.

A break in the stalemate

The contractor's site manager again wrote to the architect, demanding an explanation of the second claim's rejection. The architect conveyed the contents of the letter to the client's project manager who immediately contacted the contractor's site manager to promise action. Once again, the client's project manager had been unaware of the ongoing dispute, as had the client's QS, who discovered it in an unrelated conversation with the contractor's site manager. In two subsequent meetings, the second claim was discussed but progress was hindered by disagreements over the extent of delays.

The second intervention of the regional surveyor

The contractor's regional surveyor intervened again, writing to the client's project manager complaining about the ongoing claim. The client's project manager wrote to the client's QS expressing regret that the problem had been allowed to get this far. This prompted the consultants to meet and suggest a global settlement of all outstanding claims on the project. This was an exercise in damage-limitation designed to "*save some face and wipe the slate clean for the next phase of the project*". An increasingly concerned client pushed for a solution and the idea of the global settlement was presented at a meeting. This was met with an uncertain response because their regional surveyor, who now wanted control of the situation, had removed the decisionmaking authority of the contractor's team.

The crisis is resolved

Eventually, the client's QS and contractor's QS negotiated an acceptable formula for the valuation of the second claim and agreed on a global settlement of all outstanding claims. The second claim was formally accepted and sanctioned by the client and the contract completion date was extended.

PATTERNS OF BEHAVIOR

This creeping crisis lasted approximately 10 months and consumed a considerable amount of resources, time, and energy. In all, the crisis involved people in 29 formal meetings, 68 telephone calls, and 42 letters. There would also have been many more informal meetings that were not recorded.

The exact point at which the original earthwork support problem became a crisis is uncertain, but there is little doubt that it need not have become a crisis and that eventually, it did pose a serious threat to the viability of the project, not only in terms of delays and increased costs but also in damaged interpersonal relationships. Testimony to this was the contractor's eventual decision to replace much of their project team for the second phase of the project because relationships had deteriorated to the point where they could no longer work effectively with the consultants.

The following section discusses the main phases of behavior that emerged during this crisis and explains how and when the deterioration in relationships occurred.

Phase one

Early in the crisis there was little sense of forward momentum. Communications were characterized by a sense of opposition and confrontation. This emerged out of the contractor's determination to obtain what they saw as a valid contractual entitlement and the consultant's refusal to countenance it. This period was also highly tactical. The contractor stopped work in the area affected by the earthwork support problem and the consultants called the contractor's bluff. Tactics also played a role within the "loosely-coupled" consultant's team, as the architect appeared to distance himself from the problem and the client's QS pressured the engineer to generate alternative earthwork support systems to the contractor's. The contractor retorted with expressions of commitment to the claim and warnings of delays on site. In turn, the consultants responded with delaying tactics, and eventually, a second outright rejection of the contractor's claim. Collectively, this behavior led to a phase of increasing frustration, anxiety, and confrontation. This was evident in people's communication patterns, which were characterized by numerous factions with little intercommunication as is illustrated in Figure 6-1. In Figure 6-1, factions are circled with a line thickness equal to the average number of interactions between their members. This is indicative of their relative strength and cohesion. Lines connecting factions indicate communication routes; their thickness varies according to the frequency of communication. This is indicative of the strength of ties between factions.

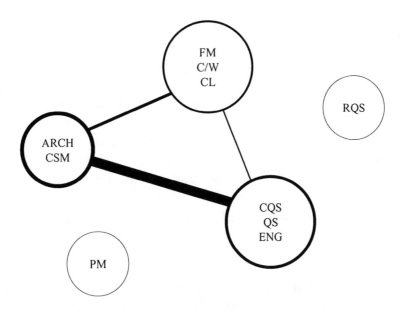

PM – Client's project manager   ENG - Engineer
CSM – Contractor's site manager   ARCH - Architect
CQS – Contractor's quantity surveyor   QS - Client's quantity surveyor
RQS – Contractor's regional surveyor   FM – Facilities manager
C/W - Clerk-of-works   CL - Client

Figure 6-1 Factional patterns during phase one.

Figure 6-1 shows three factions and two isolates. The relative isolation of the client's project manager indicates that he was excluded from negotiations surrounding the claim. His mistake was to rely on the architect as his only point of contact with the project team, thereby making himself vulnerable to the architect's vested interests, which were best served by not widely publicizing the problem. Most noticeably, the architect and client's QS were in separate factions, supporting the emerging picture of the architect's desire to distance himself from the problem. Despite their separation, they did have a healthy level of communication, although the architect primarily acted in a receiving capacity, relying on the client's QSs to coordinate a response. Indeed, the architect, being in the strongest faction with the contractor's site manager, appeared to use the client's QS's advice to perform an important bridging role between the consultants and contractor. This would have enabled him to maintain control of the situation but at the same time avoid direct implication in it. As a further point, the engineer was by far the weakest member of his faction, only having contact, in a receiving capacity, with the client's QS. This is evidence of the pressure being exerted upon him to generate alternative earthwork support solutions to that proposed by the contractor. However, his strong connection with the contractor's site manager in the architect's faction was also evidence of his sympathies with the contractor's case. This eventually led him to "leak" information, which equalized information differences between the contractor and consultants, thereby undermining the latter's bargaining position.

In terms of people's centrality to communications, there was no clear "source" or "sink" of information and therefore little sense of clear leadership during this phase. However, the architect and client's QS occupied the main gate-keeping positions and thereby, exerted considerable control over information flow. This made the crisis management process dependent on their relationship with the contractor, which was characterized by mutual suspicion and distrust. In essence, it appears that the communication structure that emerged during this initial phase contributed significantly to its inefficiency by making the crisis management process vulnerable to the negative relationships that existed among a few key individuals.

Phase two

Phase two began with a dramatic increase in forward momentum compared to phase one. The sudden involvement of the contractor's regional surveyor, an escalation of the crisis, a sudden movement toward more aggressive tactics, and a greater show of emotion brought about this change. In response, the consultant's policy of opposition and suppression in phase one was replaced by increased attention to resolving the problem. The level of opposition fell, parties were more concessionary, and there was a higher level of discussion about the contractor's claim. In essence, the regional surveyor's intervention induced a more productive and supportive phase where open discussion replaced the manipulative, coercive tactics that characterized most of phase one.

Collectively, these conditions led to a gradual decline in emotions, frustration and anxiety that was reflected in higher levels of more effective communication among project participants. This is illustrated in Figure 6-2 which shows the client's project manager in the strongest faction with the architect and client's QS. His increased involvement appears to have been a defensive response to the sudden escalation, brought about by the regional surveyor's (RQS) intervention. The engineer is now in a faction with the contractor's site manager and regional surveyor, his separation from the consultants reflecting a greater focus on resolving the problem rather than merely generating alternative earthwork support systems to the contractor's.

A further contrast to phase one was a higher level of direct contact among people and a higher level of equivalence in their personal communication networks. This indicated widespread access to similar information and a greater chance of mutual understanding of relative positions in negotiations. In phase one, conflicts of interest forced people to protect their information sources, thereby causing confusion, misunderstandings, frustration, and mistrust.

In addition to being more direct and open, communications were more centralized around specific individuals, indicating a more closely integrated and tightly knit team. By far the most central people were the contractor's site manager and the client's QS, indicating their leading role in resolving the dispute. In contrast, the architect had a relatively low centrality compared to phase one, confirming the architect's continuing desire to see the client's QS take responsibility for the problem. The information gate-keeping structure was similar to phase one but it did not adversely affect information flow because of more positive attitudes among the consultants and the contractor.

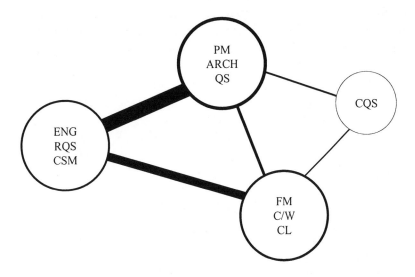

PM – Client's project manager
CSM – Contractor's site manager
CQS – Contractor's quantity surveyor
RQS – Contractor's regional surveyor
C/W - Clerk-of-works

ENG - Engineer
ARCH - Architect
QS - Client's quantity surveyor
FM – Facilities manager
CL - Client

Figure 6-2 Factional patterns during phase two.

Phase three

Phase three coincided with the contractor's second claim and was characterized by a dramatic reduction in forward momentum compared to phase two. This was largely a consequence of the architect's tactic of ignoring it, which prompted the contractor to respond with warnings of delay, threats, and eventually, an act of escalation that involved the second intervention of their regional surveyor.

This increased sense of division and confrontation was reflected in the dominance of two loosely coupled factions, one comprising the architect and contractor's site manager and the other comprising the facilities manager and clerk-of-works. This is illustrated in Figure 6-3, the latter faction being primarily concerned with the technical challenge of resolving the collapsed water main and the former with the contractor's second claim.

During this phase, information flow increasingly centered around the architect and the contractor's site manager, indicating that they were considerably more knowledgeable about the ongoing dispute than other project members. This widespread ignorance of the on-going dispute, beyond the architect and contractor's site manager, would have been exacerbated by the gate-keeping roles they played within the project's communication network. This gave them considerable control over information flow among people, making the crisis management process vulnerable to their poor relationship. In a reflection of phase one, it would seem that the communication structure that evolved among people during this phase would have played a considerable role in the lack of forward momentum and acrimony that characterized it.

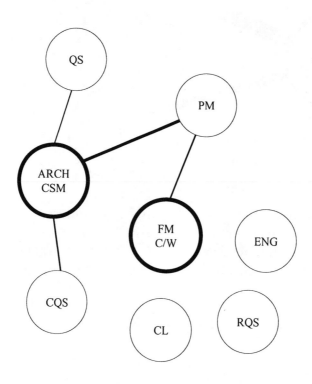

PM – Client's project manager
CSM – Contractor's site manager
CQS – Contractor's quantity surveyor
RQS – Contractor's regional surveyor
C/W - Clerk-of-works

ENG - Engineer
ARCH - Architect
QS - Client's quantity surveyor
FM – Facilities manager
CL - Client

Figure 6-3 Factional patterns during phase three.

Phase four

The final phase of behavior was characterized by a dramatic increase in forward momentum and an increasingly cooperative, compromising, and supportive atmosphere, compared to phase three. This was brought about by the second intervention of the contractor's regional surveyor, a tactic designed to resolve the stalemate surrounding the contractor's claim. There was also a greater focus upon problem resolution through open discussion and negotiation, which was reflected in calmer emotions, growing contentment and reduced rhetoric in communications. Patterns of communication were also less divided in that there was only one dominant faction, which consisted of the architect, client's QS, contractor's QS, and client's project manager. This is illustrated in Figure 6-4.

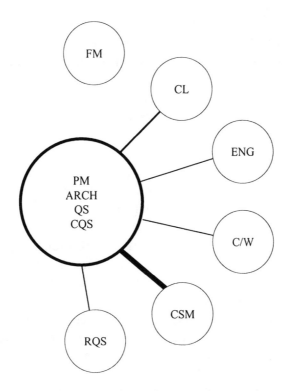

PM – Client's project manager	ENG - Engineer
CSM – Contractor's site manager	ARCH - Architect
CQS – Contractor's quantity surveyor	QS - Client's quantity surveyor
RQS – Contractor's regional surveyor	FM – Facilities manager
C/W - Clerk-of-works	CL - Client

Figure 6-4 Factional patterns during phase four.

Figure 6-4 indicates that one very tightly knit group who took control of the crisis, working closely to bring it to a conclusion, dominated the final phase of the crisis management process. While the contractor's site manager was excluded from this faction he was strongly connected to it.

The central players during this phase were the contractor's QS, client's project manager, architect, and client's QS. This reflects a breaking down of the contractor/consultant divide that had developed in phase three and an injection of consultant effort to get the problem resolved. In contrast to phase three, the architect played a far more dominant sending role, indicating that his policy of silence had ended and that he was driving the process toward a conclusion.

A particularly interesting development was the client's project manager's movement into a position of high "betweenness," which enabled him to exert greater control over the crisis management process. In essence, he presented himself as an alternative route for the contractor's communications thereby overcoming the dominating effect of the poor relationship between the contractor's site manager and architect in phase three.

Finally, in a further contrast to phase three, there was a rise in the equivalence of people's communication networks. This indicated a period of widespread communication that enabled people to construct a common understanding of the problem and thereby reach a mutually agreeable solution. Collectively, these communication patterns led to a healthy level of inter-personal communication and a far more positive period of activity than in phase three.

CONCLUSION

This conclusion uses the cyclical model of crisis management depicted in Figure 3-1 to discuss how effectively this crisis was managed.

This crisis was self-manufactured in that it grew out of a relatively simple problem that was poorly managed. The problem that evolved into this crisis had laid dormant for some time, having been caused by an error in constructing the bill of quantities. Although the contractor had been aware of the problem for some time, he delayed notification because of an inherent distrust of and a conflict of interests with the architect. The rationale was that delaying the notification until the last minute would increase the probability that the response would go in their favor.

Thus, early inefficiencies were not of monitoring, as superficially appears, but of poor communication among monitors and comparators caused by a conflict of interest. Once notified of the problem by the contractor, the consultants, acting as comparators, decided that the problem was the contractor's. Essentially, they attempted to terminate the crisis management process at the first opportunity, forcing the project team back into a monitoring mode. This series of events are illustrated in Figure 6-5.

This tactic was a protection mechanism, motivated by self-interest and designed to avoid the problem, call the contractor's bluff, test the contractor's resolve, and transfer the onus of proof back onto the contractor's shoulders. In essence, the consultants attempted to keep the problem contained within the confines of their own power base by acting as both comparator and decisionmaker. Indeed, by avoiding the need to invoke higher levels of decisionmaking authority, the consultants managed to conceal the problem from the client's project manager. To reinforce this tactic, they became inwardly oriented and imposed strong group-norms, particularly on the engineer, to construct a highly biased definition of the problem from their own perspective.

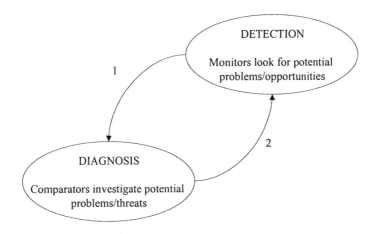

Figure 6-5 Initial attempt to terminate the crisis management process.

Faced with increasingly polarized positions, the contractor resorted to coercive power tactics by threatening the consultants with delays and an escalation of the dispute. Although this tactical escalation was successful in getting their claim recognized, the consultants made a decision that was outside their authority. While diffusing short-term tensions, long-term tensions were increased because the client's project manager, who had the necessary authority to act in a decisionmaker's capacity, refused to do so. Before sanctioning the consultant's decision, the client's project manager insisted on a reassessment of alternative earthwork support systems and in doing so, returned the crisis management process to a diagnosis stage, prolonging it and frustrating everyone concerned.

This series of events is illustrated in Figure 6-6 where the dotted lines record previous movements among the phases of the crisis management process.

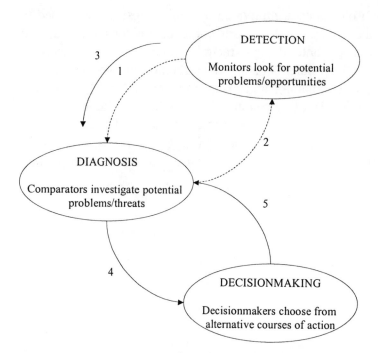

Figure 6-6 Returning the crisis management process to a diagnostic mode.

The client's project manager eventually sanctioned the claim but then, the earth bank to be supported by the disputed earthwork support system collapsed. This made the claim irrelevant and threw the crisis management process back into a diagnostic mode to resolve the new problem. This series of events is illustrated in Figure 6-7.

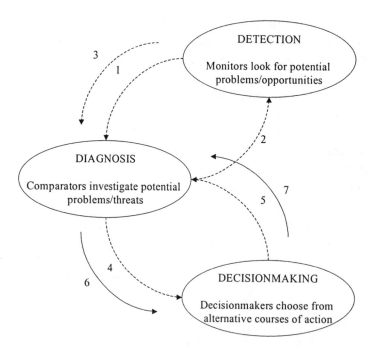

Figure 6-7 Returning the crisis management process to a diagnostic mode again.

Paradoxically, this sudden sub-crisis caused a temporary alignment of interests and increased cohesion within the project team because the contractor advocated a lower cost earthwork support system that was duly sanctioned by the consultants. This series of events is illustrated in Figure 6-8.

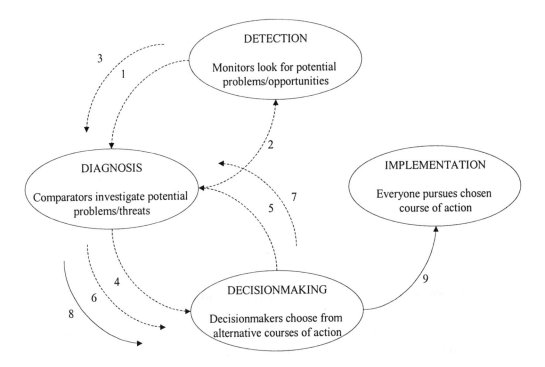

Figure 6-8 The issue and implementation of a new change-order.

Up to this point, the crisis management process as depicted in Figure 3-1 had not been a smooth cycle but one characterized by considerable inefficiency. In particular, the process appears to have been characterized by a considerable degree of repetition and procrastination in moving between the different phases of the crisis management process. Indeed, this continued because during implementation of the revised earthwork support system, the crisis management process was thrown into a second full cycle by the contractor's second claim for an extension of time. This is illustrated in Figure 6-9.

The need for a second cycle of the crisis management process was a direct consequence of delays caused by inefficiencies in the first cycle. The contractor had monitored these delays for some time, but in a reflection of the first cycle, mistrust of the consultants caused them to withhold their notification. Another similarity with the first cycle was that the subsequent diagnostic process was characterized by defensiveness on the part of the architect and a reluctance to recognize the problem. This led to a build-up of frustration and eventually a second escalation of the crisis. This initiated a new diagnostic process that involved a full consideration of the problem by all of those with vested interests in its solution.

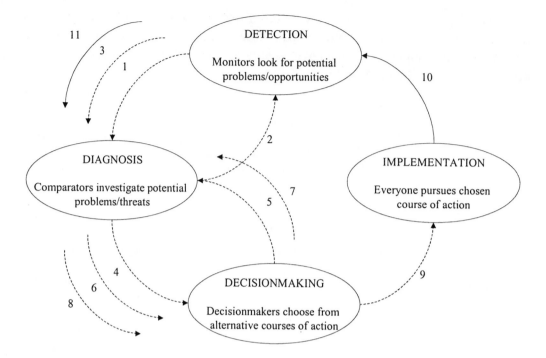

Figure 6-9 A second cycle of the crisis management process.

Eventually, after a convergence of views, the contractor's claim was granted and the project's completion date extended. These events are illustrated in Figure 6-10.

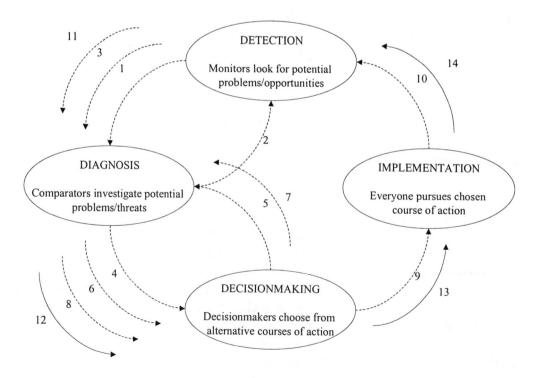

Figure 6-10 The completion of the crisis management process.

Chapter 7

Case Study Two

THE PROJECT

This project involved the construction of a new museum. The contractor considered this project of strategic importance in moving into a new geographical area and had no previous experience of working with any of the client's consultants.

THE CRISIS

A "sudden" crisis was caused by a decision to incorporate an eight-person elevator into the scheme, as construction was about to begin on site. This decision was prompted by complaints regarding the lack of provisions for disabled people in the scheme. The decision to include an eight-person elevator had significant design and programming implications, and because the discussion came late, gave the project team very little time to implement.

AN ACCOUNT OF THE CRISIS MANAGEMENT PROCESS

Keeping the project going

What made this crisis particularly difficult and frustrating for the architect was his lack of involvement in the original design process. He was not fully conversant with the designs and felt slightly resentful about rectifying someone else's mistakes. Furthermore, when the problem was highlighted, he was about to send out tender documents to main contractors. These had to be rapidly adjusted to incorporate the possibility of including an elevator in the scheme. To this end, after first failing to persuade local government inspectors to permit other means of access for the disabled, a decision was made to include a provisional sum of money in the bill to cover the cost of an eight-person elevator. The client formally authorized this provisional sum within a week but withheld the final decision about the elevator's inclusion until budgets were clarified for the next financial year.

A decision is made

After five months, the client contacted the architect to confirm the elevator's inclusion in the scheme. The main contractor was about to begin construction work on site but no progress had been made on redesigning the building. The main problem now facing the architect was the enormity of re-design work, the timeline for undertaking it and the need to get an elevator sub-contractor on board as quickly as possible.

Within a week, the architect had sent tender documents to four elevator sub-contractors and during the pre-construction meeting with the successful main contractor announced the inclusion of the elevator. After the initial shock and amusement had subsided, the prospective site manager requested the drawings. The architect pointed out that revised drawings were not yet available and that he would try to supply them as soon as possible. The architect's explanation for the lack of design information was that "*the inclusion of the lift was dependent on budget allowances and with the pressure we are under, you just can't progress with design if there is the slightest chance of it being wasted work.*"

The site manager recalled, "*I was starting on site the next week but I was afraid to start anything. It was just not knowing what was happening...we discovered later that the architect knew about the lift for some time. If we had been told straight away, we could have been looking at the program.*" While the elevator's inclusion was a surprise to the contractor, it was not a surprise to the mechanical and electrical (M&E) engineer. He had been a member of the original design team and had warned of the need for an elevator: "*They just ignored me, so I thought, 'it's your problem not mine.' Of course, it's come home to roost now hasn't it? It is my problem. They were just under time pressure and they knew that, in all probability, they would not be involved in construction and so they could just pass the problem up the line.*"

Location of elevator is decided

The day after the pre-construction meeting, the elevator's position in the building was agreed upon.

Work commences on site

As foundation work proceeded on site, the site manager had numerous informal discussions with the clerk of works about his growing concerns over the lack of information, particularly about the elevator pit position. The clerk of works conveyed these concerns to the architect and explained that, "*the site manager was having to go to a lot of trouble to work around the problem on site*".

Drawings are supplied indicating elevator pit position

One month after the elevator pit's position had been agreed upon, information about the pit size and position was faxed to the contractor. The elevator was to be positioned in the center of the building, and because there was only one point of access to and from the site, completed foundations had to be backfilled to allow excavators to track across them to get to the elevator pit location. This information had been withheld in fear of the elevator position changing and because the architect preferred to issue complete packages of information rather than *"drip-feeding"* it: *"I wanted to start as I meant to go on, by not issuing drawings until they were complete. That's all I needed was more changes."* However, both the clerk-of-works and site manager were becoming increasingly frustrated at the architect's formality and inflexibility in issuing information and pointed to the unnecessary disruption it was causing: *"[the architect] likes to do things by the book. Its good in some ways but when you need the information it causes lots of problems because we've got to keep going on site."*

The elevator sub-contractor is selected

The architect received tenders from elevator sub-contractors and entered into contractual negotiations with the lowest cost tenderer.

A continuing lack of information

The site manager continued to express his concerns about progress on site and his desperate need for revised drawings.

Structural engineering drawings provided for elevator pit area

Revised structural engineering drawings for the new elevator pit and adjacent foundations were supplied after continuous requests from the site manager and clerk-of-works. The architect continued to negotiate with the successful elevator sub-contractor but a problem arose over liquidated damages provisions in the sub-contract. This delayed formal nomination, preventing any contact between the main contractor and sub-contractor.

A second crisis

Eventually, after resolving contractual problems, the prospective elevator sub-contractor visited the site and noticed that the elevator pit excavations were too small to accommodate the elevator, which had been specified in their tender documents. The situation was made worse because the sub-contractor did not have a standard elevator to fit the pit that had been substantially completed. On the same day, the engineer and clerk-of-works became concerned about the depth of the elevator pit foundations, something that had implications for adjacent foundations.

It emerged in subsequent discussions that to enable re-design work to progress, the architect had made an inaccurate assumption about the elevator pit size before the elevator sub-contractor had been selected. All design consultants had worked on the basis of this inaccurate decision in re-designing their part of the building, and along with the contractor, now faced a huge amount of "re-re-work" to accommodate the elevator sub-contractor's standard elevator specifications and designs.

An urgent meeting was called, during which there was widespread concern about the problem. However, there was also a degree of sensitivity and sympathy for the architect's predicament and humor played a great part in diffusing interpersonal tensions. Rather than exploiting the situation, the contractor was supportive and accommodating in suggesting another temporary reorganization of the site to alleviate the immediate pressure on the architect: *"We didn't want extras, we just wanted to be able to pick up the drawings and build. We wanted a successful project as much as the architect, so we did anything we could to help. It is important to us that the project goes well and [the site manager] was able to reorganize things on site so that we only lost a few days. We didn't have a delay because we didn't want one, but if we were bloody minded we could have really made trouble."* This flexibility and goodwill impressed the architect: *"this contractor is the best I have dealt with in a long time, all on site and in their office are exceptionally pleasant people to deal with. I had a bad experience on my last job and they have re-confirmed my faith in contractors."*

Specifying a custom-made elevator

In another rapidly convened meeting, the architect, site manager and elevator sub-contractor decided to consider a custom-made elevator. The architect and sub-contractor negotiated a price that was within the original provisional sum, which meant that budget allowances were not exceeded.

Formal nomination of the elevator sub-contractor

The elevator sub-contractor had not yet been formally nominated and any involvement so far had been on the basis of goodwill. Upon formal nomination, the architect instructed the contractor to start discussing program details with the sub-contractor. Also, in response to contractor requests, the architect promised a formal "change order" to cover the costs of the elevator's inclusion. The site manager stressed his urgent need for a revised elevator specification and pit design for the new custom-made elevator.

New elevator pit designs are issued

The architect sent the elevator sub-contractor's specification, designs and contractual conditions to the contractor. The same were sent to the M&E engineer with an urgent request for amended M&E drawings. The M&E engineer recalled his annoyance: *"He doesn't seem to realize that when he changes one of his drawings we have to change ten of ours."*

Further information shortages

During the next monthly site meeting the contractor expressed concern about the lack of change-order and M&E drawings. In a tolerant and good-humoured environment, the contractor was reassured of their imminent issue. The following three weeks were characterized by numerous telephone calls and meetings between the site manager, contractor's QS, architect, and clerk-of-works about the outstanding M&E drawings that were desperately needed on site. Eventually, the M&E engineer sent his revised drawings to the architect and they were issued the day they were received. The contractor's QS expressed continuing concerns about the lack of change-order; a month later, the architect issued it: "*It would have put our minds at rest if we could have had it earlier but this is just the way [the architect] likes to work.*"

PHASES OF BEHAVIOR

The response to this sudden crisis lasted for approximately nine months and consumed a considerable amount of resources, time, and energy by involving people in 33 formal meetings, 27 telephone calls, and 61 letters. There would also have been many informal meetings that were not recorded.

In contrast to the crisis described in Chapter Six, there were no discernible phases of distinct behavior during this crisis. Throughout, the problem was one of information management rather than of handling a dispute, the pressure being focussed primarily on the architect's shoulders. Indeed, everything indicates that the architect failed to come to terms with the information demands placed on him. He operated in a reactive mode, supplying information in periodic surges rather than continuously, and he was always "chasing" the demand for information rather than leading it. This was largely a consequence of the client's procrastination in sanctioning the elevator, causing the architect to ignore the "time-window" for re-design, which occurred between tendering and construction. This missed opportunity made a complex problem a crisis by ensuring that construction activity was plagued by a lack of information. This caused a considerable degree of uncertainty, frustration, and organizational inconvenience for the contractor and tested interpersonal relationships on the project.

The project's savior was the contractor's site manager who compensated for the architect's periodic lapses of attention to information supply by increasing his persistence in requesting information. While the architect rarely responded immediately to the contractor's requests for information, he was sensitive enough to supply information in enough time to prevent any damaging accumulation of tension.

Despite these information supply problems, there was widespread commitment to resolving the crisis, and as the crisis continued, there emerged a sense of mutual support, sensitivity, and consideration among everyone affected. This was particularly evident when the pit-size discrepancy was discovered—a problem that created a potentially explosive atmosphere. However, this sub-crisis strengthened inter-personal relationships rather than damaged them because people used it as an opportunity to demonstrate their commitment to each other and to the project's success. There is little doubt that the contractor's determination to impress what was a new and potentially lucrative client was a major factor that contributed to the lack of exploitation and conflict throughout this crisis. Furthermore, the majority of people involved in this project were nearing retirement and took a very philosophical approach to the crisis, having seen it all before. This enabled them to place it in its proper perspective, and rather than panicking and creating a volatile atmosphere, they made a conscious attempt to suppress emotions, stay calm, and remain rational.

The communication patterns that emerged during this crisis were dominated by two main factions, as is illustrated in Figure 7-1.

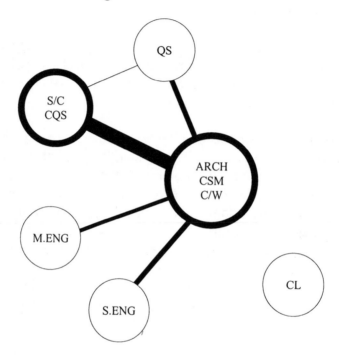

S.ENG – Structural engineer   M.ENG – M&E Engineer
CSM – Main contractor's site manager   ARCH - Architect
CQS – Main contractor's quantity surveyor QS - Client's quantity surveyor
S/C – Elevator sub-contractor   C/W - Clerk-of-works
CL - Client

Figure 7-1 Faction structure during case study two.

Figure 7-1 indicates that the strongest faction was among the site manager, architect, and clerk-of-works. Within this faction, the clerk-of-works played an important bridging role by maintaining communications and creating a "buffer" between the architect and site manager. This minimized the chances of conflict in a stressful and potentially explosive environment.

Although the architect, clerk-of-works and site manager worked closely, the architect held the most central communications position, which meant that were was no competing source of information and a high degree of consistency in understanding the nature of the crisis. Unfortunately, one of the problems with the architect's high centrality was other's dependence on him, which made the crisis management process highly vulnerable to his vested interests and to his ability to deal with the extreme pressures on him. Although there was no evidence of the architect manipulating information to serve his own interests, evidence did indicate that the architect became increasingly unable to cope with the information demands on him. His survival mechanism in coping with this information overload was to adopt an increasingly distant, formal, and inflexible managerial style, which only compounded the lack of information that was causing the pressure. The result was a considerable degree of uncertainty and frustration for the contractor. Important in alleviating this pressure before it caused conflict was the site manager's compensating role in proactively seeking information from him and reorganizing work on site, when he was unable to cope.

CONCLUSION

This conclusion uses the cyclical model of crisis management depicted in Figure 3-1, to discuss how effectively this crisis was managed.

This crisis arose out of a design-stage problem that had been ignored due to time pressures and changes in project membership between design and construction stages. The personnel change also desensitized the construction-stage design team to the potential problem meaning that the re-detection of lifts omission came late, arising not out of the organization's diligent monitoring of its environment but vice-versa. Once the problem was detected, the architect, as the comparator, was responsible for determining the extent to which it threatened the client's goals. The client's main concern was to control costs and the architect's initial response was an attempt to avoid the problem by trying to persuade local government inspectors that an elevator was not required. In essence, he tried to return the crisis management process to a monitoring mode. This series of events is illustrated in Figure 7-2.

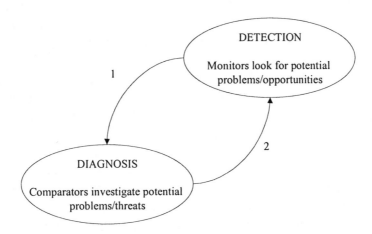

Figure 7-2 Initial attempt to terminate the crisis management process.

Recognizing the seriousness of the problem, the architect recommended that the client include a provisional sum in the bill of quantities. Ironically, while the detachment of the architect from the original design process may have adversely affected monitoring activities, it probably helped speed progress between comparator and decisionmaker because the architect was not implicated in any blame and therefore, was not defensive. These events are illustrated in Figure 7-3.

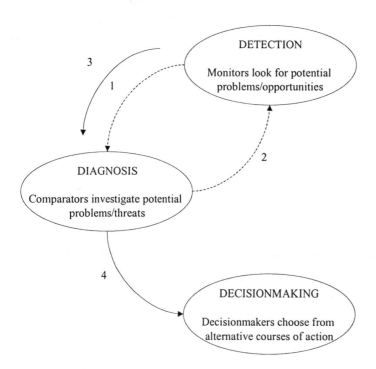

Figure 7-3 Involving the decisionmaker.

In acting as decisionmaker, the client took the advice of the architect and quickly decided to include the provisional sum. However, this was only a provisional decision and the client took an inordinate amount of time in making the final decision about whether or not to include the elevator. It is debatable whether the client was to blame for this delay, since the architect, as lead consultant, was also complacent in failing to make the client aware of the importance of making a quick and firm decision. This appeared to be motivated by the daunting amount of redesign work involved and the hope that the client's decision would be negative and the problem would go away. Whatever the reason, this delay in decisionmaking proved to be the real cause of the crisis in allowing information demand to run ahead of information supply. It was an implementation problem from which the architect never recovered, and one that forced him into an increasingly reactive mode of management. This problematical movement into the implementation phase is illustrated in Figure 7-4.

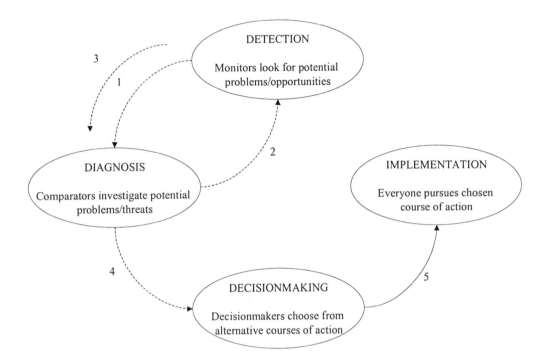

Figure 7-4 Implementation.

Problems began to "snowball" as the constant pressures to supply information caused the architect to make a hasty assumption about the size of the elevator pit. This led to an under-sized elevator pit being constructed, which launched the crisis management process into a second cycle. During the diagnostic process, the project team was lucky because the elevator sub-contractor was able to supply a custom-made elevator for the same cost as a standard elevator. In this sense, this sub-crisis did not pose a further threat to client goals and the client, as decisionmaker, was not involved for the second time. The emphasis then returned to the implementation of this revised decision. This series of events is illustrated in Figure 7-5.

During the second implementation process, the architect continued to contribute to the pressure of the situation and to the contractor's frustration through his dislike of uncertainty. This caused him to inappropriately treat a non-routine situation as routine, by not adapting his "normal" information production procedures to the extreme demands of this crisis. The main problem was that the architect continually insisted that information packages were fully complete before issuing them, which meant that information was supplied in surges rather than constantly, as preferred by the contractor. The architect also failed to recognize the contractor's information priorities by not giving special attention to particularly important items of information that were urgently needed to maintain progress on site.

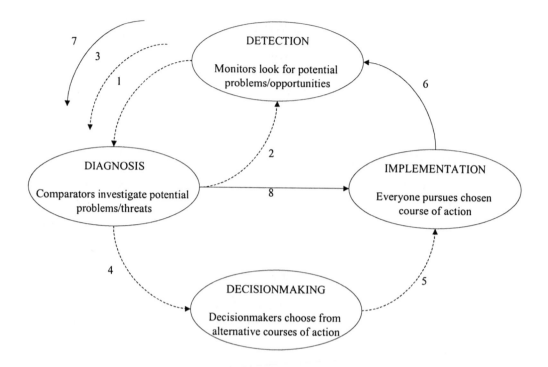

Figure 7-5 A second cycle of crisis management.

Despite the above problems, the crisis did not result in an extension of the project's completion date and the extra costs were contained within the monetary limits of the provisional sum. This considerable achievement was largely a consequence of individuals who recognized and capitalized on the opportunistic aspects of the crisis. In particular, the contractor treated it as an opportunity to demonstrate commitment to the project by exhibiting a willingness to tolerate the inconvenience of poor information supply without claiming compensation. There is little doubt that the contractor could have exploited the situation. The contractor also recognized its collective responsibility for information supply by constantly keeping the architect informed of information needs rather than merely relying on him to supply it. In this way the contractor played an important role in compensating for the architect's weaknesses and thereby helped to nurture a sense of cohesion within the project team and to generate a culture of mutual responsibility for the problem. In essence, the efficiency of this crisis management process is largely attributable to the exploitation, by all, of the opportunities presented by a potentially threatening and destructive situation.

Chapter 8

Case Study Three

THE PROJECT

This project was to build a new multi-million pound semi-conductor factory.

THE CRISIS

A "sudden" crisis arose as a result of a series of events on an adjacent site that was elevated and supported by an existing 120-year-old retaining wall. The construction of the new factory necessitated the replacement of the existing retaining wall with a new wall. The design of this new wall had been based on the loadings of an existing school that occupied the adjacent site. However, unknown to the engineer who had designed it, an extension to the school was planned and coincidentally, this extension started at the same time the retaining wall was being replaced.

The crisis began when the architect in charge of the adjacent school extension became concerned about the destabilizing effect of excavations near the base of the existing retaining wall that supported his site. He insisted that work stop until the wall's stability could be investigated.

AN ACCOUNT OF THE CRISIS MANAGEMENT PROCESS

The results of the investigation

This request initiated an intense period of communications that included several meetings on site with internal and external stakeholders in the project. For example, government health and safety inspectors and building control officers visited the site unannounced and meetings were held with the architect, site manager, and engineer from the adjacent site. Furthermore, the head master and board of governors of the school on the adjacent site had to be informed of the safety implications for children, as did others in the local community. The end result was the discovery that the engineer, unaware of the new extension to the adjacent school, had under-designed the new wall. Consequently, a new design was required. Since almost all site activities on the new factory depended on the retaining wall being completed, this had an enormous impact on construction work.

The site is shut down

The option of shutting down the site was discussed with the client, but before the shut down, the existing retaining wall had to be restabilized for safety reasons. A geologist was consulted regarding ground stability and a meeting was quickly convened on site to inspect the excavations. After the various options for supporting the excavations were discussed, it was decided that a monitoring system should be set up and that an "*exclusion zone*" of 5 meters be imposed on any work behind the existing retaining wall on the adjacent school site. In a second meeting with the client, a decision was made to keep a "skeleton-staff" on site to monitor the excavations, to keep the general public out, and to undertake minor works that did not depend on the retaining wall designs. The contractor was also asked to be ready to re-start the site at short notice.

Re-design commences

The decision to stop the site initiated a period of intense design activity, starting with the drainage system so the contractor could recommence site activities as soon as possible. As the architect recalled, "*We worked every waking hour to re-do the drainage layouts, which involved going right back to the beginning and starting again.*"

The revised drainage system is rejected

The revised drainage drawings were eventually sent to the local government building control department for inspection but were rejected because they did not comply with local building regulations. In a revealing comment about the stress induced by this sub-crisis, the architect recalled, "*everyone suffered, but unfortunately it all took its toll on [the engineer]. You imagine it, everyone standing around waiting for you. The QS was on his back to keep costs down, [the site manager] for information, and our director to restart the site.*" In the meantime, an increasingly concerned client visited the site unannounced to discuss progress with the site manager.

The revised drainage system is approved

Eventually, the revised drainage system was approved and a decision was made to restart the site. However the site manager recalled, "*they gave me the drainage details, yes, but most of it was behind the retaining wall which was not built at that time or even designed. In other words I couldn't use it.*" Indeed, during the next site meeting the site manager warned that he would "*soon be on stop again*" if information about the retaining wall was not forthcoming.

The revised retaining wall details are supplied but are incomplete

The engineer issued, in person, to the site manager, the majority of the revised retaining wall details. However, upon inspection, the site manager discovered that information about concrete mixes and reinforcement details were missing. The engineer verbally gave the site manager the missing information.

Logistical problems in constructing the new retaining wall

Excavations for the new retaining wall began, but a large pocket of soft ground was encountered and the engineer issued a verbal instruction to excavate it and fill it with concrete. Engineer's drawings continued to be supplied with incomplete information and the site manager repeatedly requested a decision from the architect about the relocation of existing overhead electrical and telegraph cables so machines could operate safely in the excavation area. In site meetings, the site manager was becoming increasingly agitated with the constant lack of information.

New problems with the engineer's drawings

The engineer continued to deliver incomplete retaining wall drawings and to visit the site to verbally redress the deficiencies the site manager discovered. For example, on one occasion, the engineer issued a verbal instruction specifying the retaining wall's finish and increasing the volume of concrete in its foundations to compensate for differences in site datum compared to those shown on the drawings. Another problem that the site manager detected was related to insufficient pipe bedding behind the retaining wall. Once again, the engineer issued a verbal instruction to increase it.

The site manager became increasingly frustrated: "*At this time I was operating on site with virtually no information. The engineer just couldn't cope; he was totally out of his depth and I was having to make the decisions on site as we went along from my experience. They were lucky I've got so much experience; imagine if the engineer's mistakes hadn't been picked up.... [The engineer] was suffering a lot of stress.... Eventually we reckon it put him in hospital.*"

The contractor raises the issue of compensation for disruption to their work

The contractor's QS contacted the client's QS to discuss compensation for disruption. Upon further investigation, the client's QS discovered that "*neither [the architect] nor [the engineer] had any idea of what had been said on site. Both of them had just been making changes and had obviously not given a second thought to their implications. I insisted that any further instructions had to be issued in writing.*" He suspected that he had been deliberately excluded because he would have restricted the engineer's and architect's ability to make spontaneous changes.

In a subsequent site meeting, there was little agreement about the verbal changes that the engineer had made on site, and therefore, about any reimbursement for disruption. With some emotion, the site manager recalled, "*We didn't agree on a lot and [the client's QS] started to say that some of the instructions hadn't been issued at all. But I had recorded every single change in my diary so I just sat there and read them out, one by one, in front of them all. We reckoned that there were about 55 that hadn't been paid.*"

The contractor submits a claim for an extension of time and loss and expense

The contractor's QS contacted the client's QS to warn of the continuing site disruption and served a formal claim for extension of time and loss and expense. The response of the client's QS was that "*all instructions had already been valued through normal site valuations and that any payment had to be formally justified in writing.*" The contractor was bluntly informed that "*there was no more money in the pot*" and an increasingly acrimonious atmosphere developed between the contractor and consultants.

Concerns continued to be expressed about on-going deficiencies in the engineer's drawings and schedules.

A sudden safety risk

Heavy rain caused the site manager to become concerned about the safety of the existing retaining wall, parts of which were still remaining. In response, the site manager refused to put men in that area and after a number of urgent meetings on site, the geologist was called back to advise on its stability. A decision was made to move the new retaining walls forward so excavations would not undermine the old retaining wall further.

More re-design

Moving the new retaining walls forward meant re-positioning parts of the new building, a second redesign of the drainage system, and further delays on site. Furthermore, when the new drawings were issued, the site manager detected a lack of reinforcement in some parts of the retaining wall. Unable to contact the engineer, a decision was made on site between the clerk-of-works and site manager, to increase the amount of reinforcement in the wall and to cut and bend bars on site to prevent further delays. The site manager continued to detect discrepancies in subsequent engineer's drawings, and to resolve them; a meeting was arranged on site with the engineer. The engineer did not show up. The comment in the site manager's diary was "*Waited until late but [engineer] did not arrive. [Engineer] on holidays. I hope he has arranged the information!*" The site manager recalled that "*I felt sorry for [the engineer] but I had to keep the site going and even though I did pressure him I could have been a lot harder. By this time though, he just couldn't cope and his health was suffering.*"

The contractor re-serves the claim

The contractor re-served the claim but it was rejected the same day on the basis that it was full of unsubstantiated claims and that the contractor had contributed to delays by a lack of diligence. Numerous meetings were held on site to resolve continuing drawing discrepancies.

Trying to resolve the disputed claim

The client requested an update from the architect and unexpectedly visited the site on more than one occasion. The next site meeting was very acrimonious, the client's QS re-stating his reasons for rejecting the claim and the site manager, taking personally the accusation that he lacked diligence. In response, he insisted on reading out his diary entries one by one, asking the architect and engineer to confirm them immediately. Most instructions were confirmed, and with some consternation, the client's QS asked them why he had not been informed of these changes. The site manager recalled, "*I knew I would make [the architect] and [the engineer] feel uncomfortable but unless I prompted them they just sat there in silence. They weren't going to "drop themselves in it" were they?*"

In response, the client's QS suggested that there was no proof of the work having been done, causing the site manager to angrily ask them whether they were calling him "*a liar.*" The client's QS then produced some photographs of the retaining wall drainage and argued that there was no evidence to show that certain materials, included in the contractor's claim, had been installed. The site manager disputed this and the meeting ended with no tangible progress.

Trying to resolve the disputed claim

Meetings continued to be held on site to resolve drawing discrepancies and to discuss the contractor's claim. One critical meeting lasted five hours, during which each of the contractor's 55 "heads of claim" were discussed in turn. The meeting included open accusations of unprofessionalism directed at the client's QS. These accusations had been prompted by the client's QS's suggestion to the contractor's QS that continual pursuance of their claim could jeopardize future contracts. As on previous occasions, this meeting ended with little sense of movement in relative positions.

The end of the dispute

The suggestion that future contracts would be jeopardized stimulated the contractor to escalate the dispute. This involved direct contact between the client and the contractor's managing director and resulted in the immediate and full granting of the claim.

PATTERNS OF BEHAVIOR

This "sudden" crisis lasted approximately 10 months and consumed a considerable amount of resources, time, and energy by involving people in 58 formal meetings, 67 telephone conversations, and 32 letters. There would also have been numerous informal meetings that were not recorded. The following section describes the main phases of behavior that emerged during this crisis.

Phase one

The start of this crisis was characterized by a complete neglect of financial issues because solving technical and organizational problems took priority in ensuring the immediate survival of the project in terms of physical progress on site. This may have had a role to play in ensuring that it was a predominantly positive phase with a strong sense of forward momentum. Indeed, this was also reflected in people's communication patterns, which were characterized by only one dominant faction and illustrates effective communication across the traditional consultant-contractor divide and a sense of collective responsibility for dealing with the problem. This is illustrated in Figure 8-1.

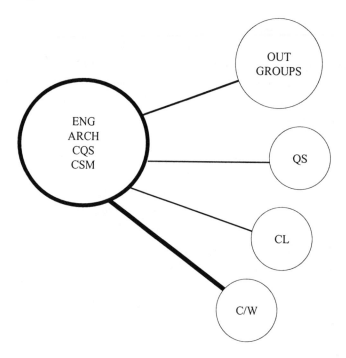

ENG – Engineer   ARCH - Architect
CSM – Main contractor's site manager   QS - Client's quantity surveyor
CQS – Main contractor's quantity surveyor   CL - Client
C/W – Clerk-of-works

Figure 8-1 Factional structure for phase one.

In terms of people's centrality to information flow, the architect and site manager were the greatest senders of information, which indicated the leading role they played in driving this phase forward. The client's QS occupied a relatively peripheral position, indicating the low priority afforded to financial issues during this initial period of activity. People were also able to communicate directly with each other, indicating a tightly knit communication structure where ideas were transferred rapidly with minimal distortion. This resulted in a widespread mutual understanding of the problem. The architect and site manager were most commonly the first point of contact for people and occupied the main gate-keeping positions within the communication network. However, while they exercised control over information flows, alternative routes among people prevented them from dominating communications. This meant that communications were not vulnerable to the attitudes and perceptions of a few people—a potential problem in the early phases of a crisis when people are likely to be at their most defensive and when the largest amount of information is generated. A further positive characteristic of communication patterns during this phase was the high equivalence in people's personal communication networks, which indicated that there would have been a common understanding of the crisis between everyone affected.

Phase two

Phase two started with the recommencement of work on site in response to growing pressure from the client and contractor. However, it quickly became apparent that the decision to restart had been premature and based on an underestimation of the contractor's information needs and an insufficient reservoir of information to fuel continued progress on site. Consequently, this phase was characterized by a reduction in forward momentum relative to phase one and by the site manager's growing sense of frustration. The main problem was related to omissions and errors in the designs supplied to the contractor, which meant that each injection of information brought with it an equivalent injection of uncertainty. This, in turn, generated a further demand for information, which drew the engineer into a spiralling cycle of uncertainty and pressure from which he found it increasingly difficult to escape. He had to chase the demand for information rather than lead it and to alleviate the pressures associated with this reactive style of management, he increasingly relied on the site manager to detect problems and identify the information needed to deal with them. In this sense, the site manager played a crucial role in supplying information.

While levels of discussion remained similar to phase one, the use of informal, verbal instructions increased substantially as a means for supplying information. This was the main mechanism used by the engineer and architect in attempting to cope with an increasingly out-of-control situation on site and to re-establish their command over information supply. These accumulating problems were reflected in the dramatic reduction in overall levels of communication compared to phase one. This is illustrated in Figure 8-2.

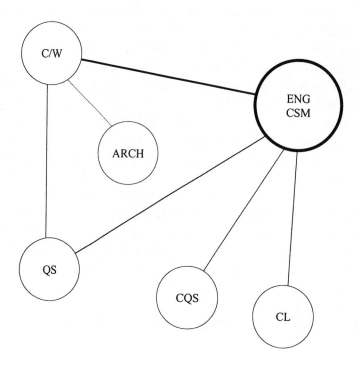

ENG – Engineer   ARCH - Architect
CSM – Main contractor's site manager   QS - Client's quantity surveyor
CQS – Main contractor's quantity surveyor   CL - Client
C/W – Clerk-of-works

Figure 8-2 Factional structure in phase two.

The existence of only one dominant faction between the engineer and the site manager indicates a dramatic focusing of communications compared to phase one, where there was a higher level of overall involvement in the crisis. This pattern was largely precipitated by the site manager who became increasingly proactive in generating his own information supply in the face of an engineer who seemed unable to cope with the pressures put on him. Indeed, during this phase of behavior, the engineer increasingly relied on the site manager's abilities to detect and notify problems on site, responding with numerous verbal instructions to change, ad-hoc, various aspects of the designs.

In terms of gate-keeping roles, the site manager became an important channel for information flow, which made the whole crisis management process vulnerable to his vested interests and personal abilities. However, the site manager was capable, experienced, and cooperative, and therefore this communication pattern had a positive influence. In contrast, the engineer's and architect's gate-keeping position between the contractor and the client's QS had a damaging effect because they used it to their own advantage to make spontaneous and informal decisions without the knowledge of the client's QS who would have undoubtedly constrained their actions. This behavior had a negative effect by isolating the client's QS from the process, which caused considerable problems in the third and final phase of behavior.

Phase three

The serving of the contractors claim initiated a third distinct phase of behavior—one that produced a dramatic focus on financial issues which, until that point, had been almost completely neglected. While a focus on technical and organizational issues had been necessary to maintain progress on site, difficulties in resolving financial issues indicated that their neglect had led to the development of very different perceptions about the costs of disruption. The consequence was an increase in emotions and a further reduction in forward momentum compared to phase two.

The client's QS became particularly frustrated and emotional about his exclusion from the design changes that had been made spontaneously on site. This placed him in a disadvantaged and vulnerable negotiating position with a relatively well-informed contractor. His response was to become defensive and avoid negotiations, which resulted in an increasingly emotional and acrimonious environment that eventually forced the contractor to escalate the dispute by involving the managing director.

This pattern of events is reflected in the factional structure that characterized this phase and that Figure 8-3 illustrates. For example, considering the focus upon financial issues, it is evident that the client's QS and contractor's QS communicated relatively infrequently. Furthermore, the divide between the contractor and the consultants seemed to strengthen compared to phase two. Finally, the architect and the client's QS worked more closely than in previous phases as the latter tried to identify the informal design changes that had been made during phase two, and from which he had been excluded. Essentially, this was an attempt to gather information that would equalize the information differences between himself and the contractor's QS, which were placing him in a disadvantaged negotiating position.

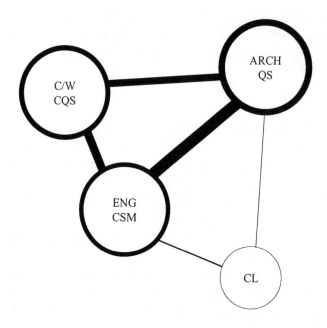

ENG – Engineer   ARCH - Architect
CSM – Main contractor's site manager   QS - Client's quantity surveyor
CQS – Main contractor's quantity surveyor   CL - Client
C/W – Clerk-of-works

Figure 8-3 Factional patterns in phase three.

CONCLUSION

This conclusion uses the cyclical model of crisis management depicted in Figure 3-1 to discuss how effectively this crisis was managed.

The origin of this crisis was the decision to extend a school building on an adjacent site, which, in turn, influenced the design of a new retaining wall. The early warning signs were missed because the crisis was not detected until some time after construction had commenced on both projects. While the engineer who designed the retaining wall could not have been expected to know of the parallel plans for an extension on an adjacent site, once its construction was started, its implications for the retaining wall's design should have been obvious. It would appear that the monitoring activities of this project organization were deficient.

Although the project organization was insensitive to the potential problem, once detected, the diagnostic process to evaluate its implications was characterized by a high level of efficiency and cooperation, resulting in a recommendation to maintain site progress in parallel with redesign. The seriousness of the crisis meant that only the client had the authority to make such a decision and this was quickly forthcoming. The implementation of this decision began immediately. This series of events is illustrated in Figure 8-4.

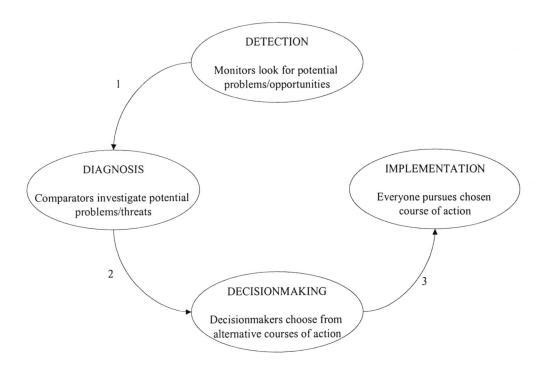

Figure 8-4 Efficient diagnosis and decisionmaking after delayed detection.

During the implementation of this decision, feedback from the site manager resulted in the detection of a further problem; delays were accumulating on site because information supply could not keep up with demand. This initiated a second full cycle of the crisis management process that began with an assessment of the new problem and the appropriate response. The process produced a recommendation to temporarily stop the site to enable design to progress to a point where it was possible to maintain construction activity on site. Once again the magnitude of this recommendation meant that it had to be referred to the client for a decision; the recommendation was duly sanctioned. The implementation of this decision meant that the site stopped, which allowed the design team to get ahead of the construction team. This series of events is illustrated in Figure 8-5.

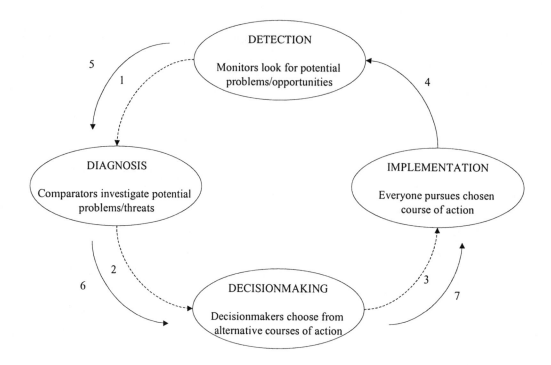

Figure 8-5 Another cycle of the crisis management process.

During the monitoring of this second period of design activity, it was determined that people had different perceptions of how long this process would take. Feedback from the client and site manager indicated increasing frustration with the delays that were accumulating on site.

As a result, they exerted considerable pressure to restart construction, which eventually resulted in a decision to do so, throwing the crisis management process into a third cycle of activity. This series of events is depicted in Figure 8-6.

The state of information supply was assessed when the decision was made to restart the site. Almost immediately, feedback from the site manager indicated that the demand for information was running ahead of its supply once again. During the stoppage, the design team had not built up a sufficient reservoir of information to support construction and they had succumbed too early to the pressures to re-start the site. Once again, information demand ran ahead of its supply and the designs increasingly became historical documents rather than the forward-planning documents they were meant to be. The pressures and stresses that this exerted on the architect and engineer led to fundamental errors and the further re-work associated with correcting them led the design team into a spiralling cycle of pressure, stress, and error from which it was increasingly difficult to escape.

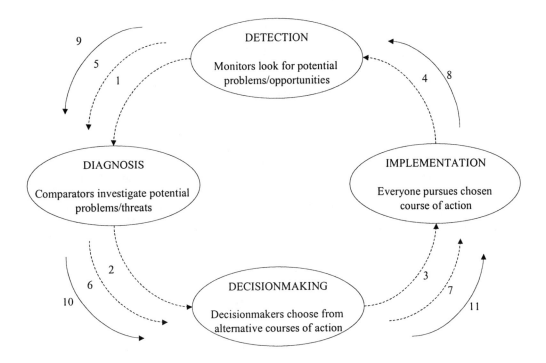

Figure 8-6 A third cycle of activity.

In an attempt to re-establish control, the engineer and architect were forced into an increasingly reactive and informal mode of management, issuing design changes spontaneously on site. While this speeded up information supply, it also created an increasingly disjointed organization. In particular, the client's QS became detached from the crisis management process and ignorant of the changes being made. Some evidence suggests that the architect and engineer deliberately filtered information to the client's QS about their design changes because he may have reduced their flexibility in issuing the verbal instructions that ensured their survival. In essence, the crisis seems to have created a conflict of interest between those responsible for the protection of different project goals (costs—the QS, and time—the architect and engineer) which prevented some parties from monitoring problems on site. In particular, the insulation of the client's QS from the site manager's feedback about continual problems on site resulted in a complete loss of financial control. However, the site manager eventually bypassed the engineer and architect to notify the client's QS of the problem, which initiated a fourth full cycle of the crisis management process which was specifically concerned with financial issues. The above series of events is depicted in Figure 8-7.

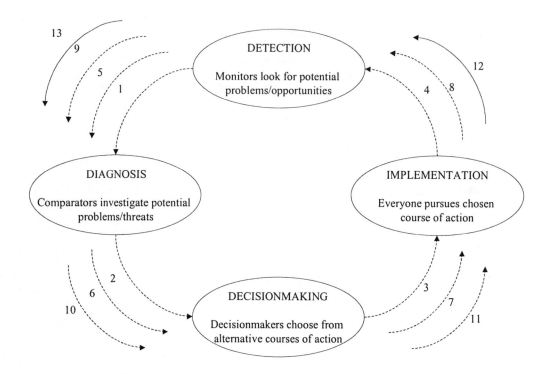

Figure 8-7 A fourth cycle of activity.

The diagnostic process was characterized by a series of acrimonious disputes between the client's QS and contractor's QS. These were a direct result of the client's QS's exclusion from the informal changes that had been made on site by the architect and engineer. This disadvantaged negotiating position forced him into a defensive mode of management. Indeed, the client's QS was never able to accumulate enough information to compete with the contractor and tactics became increasingly offensive on both sides until the dispute escalated to the point where it had to be resolved at managing director level. This is depicted in Figure 8-8.

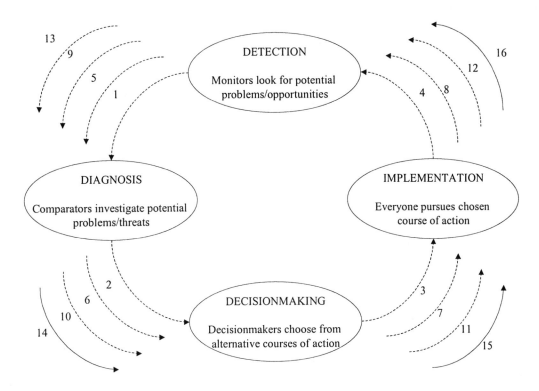

Figure 8-8 An end to the crisis.

Chapter 9

Case Study Four

THE PROJECT

This project involved the construction of a new office complex that was centered around a courtyard.

THE CRISIS

A creeping crisis began when half way through construction, a local brick paviour manufacturer of clay pavings (Company X) complained to the client that local producers did not feature in the project specifications. Although the project was well advanced, the specification was changed to incorporate Company X's clay paving. However, quality control problems quickly arose, causing serious delays and a dispute began between the contractor and architect over financial responsibility.

AN ACCOUNT OF THE CRISIS MANAGEMENT PROCESS

During the decision to change the project's specification, a discrepancy was discovered between what the contractor had priced in the bill of quantities and what the architect had originally specified. The architect argued that he had originally specified clay pavings and that the change of specification would merely involve a change of manufacturer. However, the bill of quantities had only specified concrete pavings, meaning that a change to clay would entitle the contractor to extra money.

A rift develops within the project team

This discrepancy arose from a performance specification system that allowed the architect to name three possible manufacturers from which tendering contractors could choose. However, the architect had confused matters by naming one company that only manufactured clay products, one that only produced concrete products, and one that was bankrupt. The contractor chose the cheapest option to win the bid.

The client's QS felt blamed by the architect, saying, *"The architect intimated that we were at fault for misinterpreting the specification which, he maintained, specified clay as a possible option.... The architect originally wanted clay because apparently it looked really good on his last job. But there wasn't any money in the budget so he put clay in, as an option, hoping that the contractor would go for clay. In other words he wanted clay for the price of concrete."*

Trying to reduce the extra costs of clay over concrete

The basis of the contractor's claim for extra money in using clay pavings over concrete pavings was the extra laying time (due to higher dimensional variability in clay products than in concrete products) and higher material prices. To test this argument, the architect asked the site manager to build some sample panels using the clay pavings to demonstrate the dimensional tolerance problems. He also asked Company X for their response. Company X indicated that the tolerances of clay pavings were the same as concrete and that there should be no extra laying costs. They also offered a discount that lowered their price to that of concrete.

A decision is made

After inspecting the sample panels on site, the architect opted for clay. Work quickly commenced on site because the decision had been delayed by the investigations. The contractor formally requested a change-order to cover this specification change but the architect refused, pointing to the discount offered by the manufacturer to cover the extra laying and material costs. The contractor disagreed that the discount was enough to cover the increased costs and the architect formally requested a written explanation.

Problems arise on site

A change-order was eventually issued to cover the extra costs of clay over concrete. However, the estimation of the extra costs quickly became inadequate as serious problems begin to arise on site due to high variability in paving sizes and colors. These unexpectedly wide variations had not emerged within the sample panel because of its relatively small area. The architect decided to continue with the laying.

The contractor requests a change back to concrete pavings

In addition to laying problems, the contractor began to experience supply problems and consequently, accumulating delays. The increasingly frustrated contractor formally requested a change back to concrete pavings.

The contractor communicates an intention to serve a formal claim

Company X was invited to a formal site meeting to discuss the laying problems on site. They offered to hand-pick the clay pavings, and if problems continued, to lay them themselves. Problems continued and after Company X's own team failed to cure them, the contractor wrote to the architect expressing concerns about the continuing delays and their intention to claim an extension of time.

Again, they strongly requested that the architect revert to concrete. The architect did not take the advice.

A decision is made to revert back to concrete

Eventually the laying problems became so bad that the clerk-of-works called a meeting on site to inspect the work: "*I could just see it getting worse and worse. Someone had to make a decision! The architect didn't want to because he did not want to implicate himself in the problem and the contractor didn't want to for the same reason.*" In this meeting, it was agreed that the quality of the clay pavings was unsatisfactory and a decision was made to revert back to concrete pavings. The contractor was asked to select some concrete pavings and to lay sample panels. Eventually, the architect decided on a colored concrete paviour that resembled clay.

The contractor formally serves a claim

The contractor lifted the clay pavings that had been laid and began replacing them with colored concrete pavings. The contractor also served a claim for the aborted work that was associated with the original decision to change to clay pavings. The contractor also requested another change-order to cover the latest change from clay to a colored concrete, pointing out that colored concrete cost more than the non-coloured concrete, which had originally been specified and priced by the contractor in the bill of quantities.

The claim is discussed

After a prolonged period of silence, the architect requested the client's QS's advice, which was to grant the claim. However, the architect formally rejected the contractor's claim and their request for another change-order, arguing that the original performance specification system made the contractor responsible for the choice of manufacturer. The architect argued that since the manufacturer caused the problems, it was a contractor's risk. He also maintained that the contractor caused the supply problems by being late in providing a delivery schedule to the manufacturer. Finally, he argued that the reversion back to concrete, particularly a colored concrete, had been the contractor's choice and that the architect had merely agreed with that recommendation.

This initiated a heated exchange where the contractor responded with the assertion that they had been guided to choose Company X and that without the original decision to convert to clay, there would have been no problems.

Negotiations break down

Both the architect and contractor considered suing the manufacturer. At the same time, the client's QS was informally discussing the claim with the contractor's QS, with whom he still had considerable sympathy and a good relationship from a previous project. Despite making some progress toward a compromise solution, the client's QS recalled "*In the end I had to stop talking to [the contractor's quantity surveyor] because I got the feeling that [the architect] suspected I had sympathies with the contractor and at the end of the day, my loyalties must be to the client.*" This further frustrated the contractor's QS, who recalled, "W*e just couldn't get any sense out of [the architect] but you could have a sensible conversation with [the client's quantity surveyor].*"

The dispute is resolved

The architect became suspicious of the client's QS's allegiance and took control of negotiations with the contractor's QS. Using the informal progress that had been made by the client's QS, the architect agreed to consider the increased costs for laying clay pavings and to give an extension of time, but only under the "head" of inclement weather. Within one month, a deal was struck and the contractor was given an extension of time for inclement weather and reimbursed approximately one-third of his original financial claim by using rates in the original bill of quantities. The architect also issued a change-order to revert back to colored concrete from clay, but added a written note that this was merely a confirmation of the contractor's decision. No extra costs were paid for this latter change.

PHASES OF BEHAVIOR

This creeping crisis lasted six months and consumed a considerable amount of resources, time, and energy by involving people in 25 formal meetings, 22 telephone calls, and 39 letters. There would also have been numerous informal meetings that were not recorded. The following section describes the main phases of behavior that emerged during this crisis.

Phase one

This phase started with open discussion and a strong sense of forward momentum as people investigated the feasibility of changing from concrete to clay pavings. However, soon after the decision to change to clay was made, laying problems began to emerge on site and a financial dispute developed between the contractor and architect, indicating that the decision to revert to clay was made before everyone had a complete understanding of organizational and financial issues. The result was a loss of forward momentum and a period when frustration, growing anxiety, and threats figured prominently in people's communications. The uncertainty of the architect's procrastinations in deciding to revert back to concrete contributed to further decline. The factional structure during this phase is illustrated in Figure 9-1.

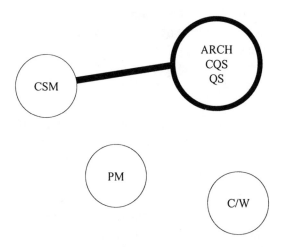

ENG – Engineer   ARCH - Architect
CSM – Main contractor's site manager   QS - Client's quantity surveyor
CQS – Main contractor's quantity surveyor   CL - Client
C/W – Clerk-of-works

Figure 9-1 Factional patterns in phase one.

Figure 9-1 illustrates that there was only one faction in this initial phase, and it consisted of the architect, contractor's QS, and client's QS, the strongest members being the two quantity surveyors. The heavy involvement of the quantity surveyors reflects the focus on financial issues in making the original decision. In contrast, the relative exclusion of the site manager reflects the lower importance attributed to the organizational issue, which eventually delayed the project and increased its cost.

In terms of people's centrality, the architect was the main source of information and drove the process forward. This reflected his enthusiasm for using clay rather than concrete pavings, his autocratic management style, and unwillingness to compromise. However, the client's QS received the most information and was clearly more popular as a point of contact for the contractor. Furthermore, like the contractor, the client's QS was being increasingly marginalized by the architect's policy of distancing himself from the laying problems that began to plague the site. Thus, while the architect was trying to maintain control by being dominant, he undermined his own position by increasing the client's QS's receptivity to the contractor. Furthermore, he drove the client's QS and contractor's QS "underground" and lost contact with their informal negotiations.

In terms of gate-keeping roles, there was no dominant person, although the architect tried to control information flow. His failure to do so was a result of the client's QS who presented himself as an alternative and much more pleasant communication route for the contractor's QS. Although he had little power to make decisions, his role was invaluable in diffusing tensions within the project team. The client's QS's actions also made him an important bridge between the architect and contractor, helping equalize information differences and avoid misunderstandings that could have further escalated the crisis.

Phase two

This phase of behavior began when the clerk-of-works called an urgent on-site meeting to inspect the worsening laying problems. The meeting lasted for a relatively short period and was characterized by a sudden surge in forward momentum, a greater willingness to face up to the increasing organizational problems on site, to investigate and discuss potential solutions, and to resolve differences of opinion. The architect, in particular, showed increased decisiveness and a willingness to take responsibility for the problem, which was in contrast to the previous phase when his energies were primarily directed toward avoiding responsibility. Consequently, a far greater emphasis on open discussion and informality emerged. This is reflected in the factional structure that emerged during this phase, and that is illustrated in Figure 9-2.

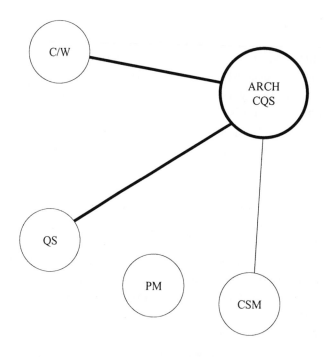

ENG – Engineer   ARCH - Architect
CSM – Main contractor's site manager   QS - Client's quantity surveyor
CQS – Main contractor's quantity surveyor   CL - Client
C/W – Clerk-of-works

Figure 9-2. Factional structure in phase two.

In this second phase, the architect and the contractor's QS formed the only faction, the client's QS was connected, but only in a receiving capacity. This reflects the sudden attempt of the architect to regain control of the process in response to the clerk-of-works' intervention. During this period, the architect assumed the role of negotiator with the contractor—a role that had hitherto been performed informally by the client's QS.

In terms of centrality, the architect and contractor's QS became dominant in both a sending and receiving capacity. This is more evidence of a new cooperative phase with a good level of communication across the contractor/consultant divide.

The end of this short-lived but positive phase of behavior coincided with the decision to abandon clay pavings and revert back to colored concrete.

Phase three

This phase was characterized by an emphasis on the financial and organizational issues that had not been fully resolved in the previous phases. However, during much of this phase there continued to be considerable disagreement over risk distribution patterns, which was responsible for a further reduction in forward momentum compared to phases one and two. A sense of forward momentum returned only when the client's QS informally intervened by discussing the claim with the contractor and when the architect assumed control to strike a deal. The factional structure that arose during this final phase is illustrated in Figure 9-3.

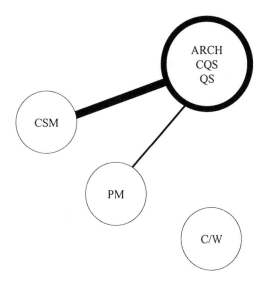

ENG – Engineer   ARCH - Architect
CSM – Main contractor's site manager   QS - Client's quantity surveyor
CQS – Main contractor's quantity surveyor   CL - Client
C/W – Clerk-of-works

Figure 9-3 Factional structure in phase three.

The factional pattern in Figure 9-3 is similar to that which arose in phase one and consists of one major faction comprising the architect, the contractor's QS, and the client's QS. The architect is the weakest member.

As in phase one, the architect refused to countenance the contractor's claim, withdrawing from the strong leadership role he played in phase two and returning to the distant role he played in phase one. Although structure was tightened around the architect compared to phase two, the majority of interactions with the architect were obstructive and prevented progress. In this sense, while the architect might have been physically close to others he was emotionally distant.

Like the factional patterns, the centrality patterns were also similar to phase one when the architect tried to maintain control by being a dominant source and gatekeeper of information supply but, in fact, lost control to the client's QS who became an alternative communication route for the contractor. Like phase one, such communications were largely informal, yet they enabled the client's QS to diffuse potential frustrations and to develop a good understanding of the contractor's needs. Ultimately, this resulted in important advances toward a compromise solution with the contractor that provided the basis for the final settlement of the dispute.

CONCLUSION

This conclusion uses the cyclical model of crisis management depicted in Figure 3-2 to discuss how effectively the crisis was managed.

The letter from Company X that precipitated this crisis arose as a result of the environment monitoring the project rather than vice versa. The responsibility to deal with the letter fell upon the architect who played both the comparator's and decisionmaker's role in assessing whether it merited a response and if so, what that change should be. A decision to change to clay pavings was made but problems soon developed in its implementation as feedback from the site indicated accumulating delays. Furthermore, a dispute over costs developed and it became evident that the original decision to convert to clay pavings had been made prematurely, on the basis of an insufficient examination of financial and organizational issues. This series of events is depicted in Figure 9-4.

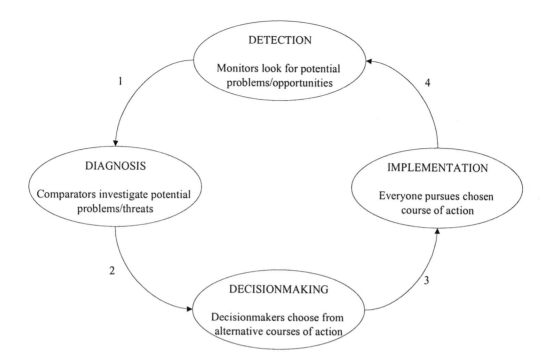

Figure 9-4 The first cycle of the crisis management process.

In responding to the feedback from site level about the laying problems, the architect procrastinated, hoping they would be resolved at site level. Furthermore, in the architect's opinion, these problems were the contractor's risk, and any intervention would implicate him. On the other hand the contractor considered it the architect's responsibility and continued laying the low-quality pavings on site. This confusion of responsibility and the fear of admitting fault, led to accumulating losses and delays that increased each party's reluctance to take the initiative to deal with the problem. Eventually, the clerk of works intervened, forcing the architect to face up to the problem and to decide to lift the clay pavings that had been laid and to replace them with a colored concrete paviour. The crisis management process was thrown into a second full cycle, the implementation of which went smoothly. This is illustrated in Figure 9-5.

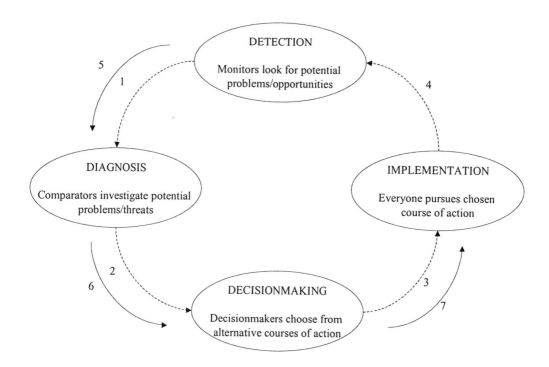

Figure 9-5 A second cycle of the crisis management process.

With the resolution of the laying problems on site, people turned their attention to the unresolved problem of responsibility for the delays and costs that had arisen as a result of the stalemate in the first cycle of the crisis management process. This issue was brought to a head by the serving of the contractor's claim, which forced the crisis management process into a third cycle of activity. The diagnosis phase of this third cycle was prolonged and confrontational because of the architect's and contractor's diametrically opposed views about who was responsible for the laying problems that had occurred on site in the first cycle. Indeed, agreement on a diagnosis of the problem was only reached because of the informal efforts of the client's QS, which played an important bridging role between the architect and contractor. This kept communications alive and prevented the dispute escalating further. This series of events is illustrated in Figure 9-6.

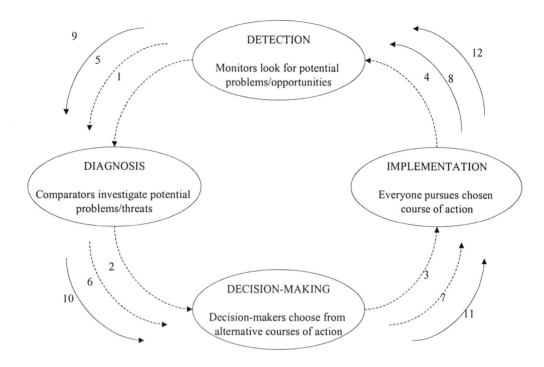

Figure 9-6 A third cycle of the crisis-management process.

Chapter 10

Lessons for Crisis Managers

This chapter compares the case studies to identify practical lessons for effectively dealing with construction crises. These lessons are highlighted in boxes throughout the text.

INTRODUCTION

The main characteristics of each case study are illustrated in Table 10-1.

Table 10-1 Case study characteristics.

Case study.	Description of crisis.
1. (Chapter 6) Earthwork support	**Creeping crisis** - A bill of quantities discrepancy relating to an earthwork support problem led to a claim for extra money and an acrimonious dispute. Procrastinations in settling it incurred considerable costs and delays and severely damaged project relationships.
2. (Chapter 7) Lift	**Sudden crisis** - Procrastinations in incorporating a new lift leads to information management problems which caused damaging stresses within the project team. These problems, in turn, damaged communications and caused significant organizational problems on site.
3. (Chapter 8) Retaining wall	**Sudden crisis** - It was discovered that a new retaining wall was under-designed. The site was shut-down but re-started too soon. This caused information management problems and stresses which caused serious delays and costs escalations on site. An acrimonious dispute also developed because of the consultants' refusal to grant the contractor's claim. Procrastinations led to the accumulation of further costs and delays and to the breakdown of project relationships.
4. (Chapter 9) Clay pavings	**Creeping crisis** - A rushed change in specification to clay pavings caused organizational problems on site and a dispute over responsibility which damaged project relationships.

A superficial inspection of Table 10-1 suggests that each crisis was unique. However, deeper investigation reveals important similarities and differences.

Built-in problems

At the most basic level, all four crises were the direct result of errors made during pre-contract stages. In each case, the seeds of potential crises were sown into the project from an early stage and were hidden by subsequent activities, only to be revealed by a "trigger event" such as the earthwork support problem in the leisure center project and the complaints of an architect from an adjacent site on the factory project.

Lessons:

- **Don't be complacent when things are going well.** Be vigilant to the early warning signs of the inevitable problems built-into projects.
- **Simple problems grow over time.** Don't suppress potential problems hoping they will go away, and don't "pass-the-buck" to people further "up-stream."
- **Keep project teams together for the life of a project.** This will prevent potential problems being lost in time.
- **If you are new to a project, investigate its history.** This will reveal potential problems that are hidden in the past and obscured by subsequent events.

Creeping and sudden crises

Occasionally, trigger events occur early enough to allow the project team to prevent a crisis. This happened in case studies one and four, but the chance was not taken and the project teams manufactured their own creeping crises in the way they responded. On other occasions, the trigger event occurs late in a crisis' development and the result is a far more potent and sudden crisis, as illustrated in case studies two and three.

The main problem in the sudden crises revolved around information management and, in turn, the pressures and stresses associated with this. In contrast, the creeping crises were constantly plagued by problems relating to procrastination, indecision, escalation, and by the management of the resultant conflict. Further differences emerge when one considers communication throughout each type of crisis. For example, in the sudden crises, early efforts at resolution were primarily focused on technical and organization issues and were largely constructive in nature. That is, the suddenness of the crises appeared to make the immediate survival of the organization an initial priority over the pursuit of self-interest and forced people together. Problems relating to the settlement of financial responsibilities tended to be postponed and be attended to later in the crisis. However, when attention was eventually turned to financial issues, the delay appears to have allowed differences of interests to have become stronger and more clearly defined, resulting in particularly acrimonious conflicts.

In contrast, the creeping crises were primarily concerned with the resolution of financial responsibilities, particularly early on. Technical and organizational issues were a concern, but to a far lesser extent and typically postponed until later in the crises. In contrast to sudden crises, creeping crises gave people time to think about their differences and to formulate complex tactical strategies to force resource redistributions in their favor. This resulted in procrastination in making decisions, prolonged conflict episodes, and considerable periods of uncertainty and frustration for everyone involved.

Lessons:

- *Creeping and sudden crises present different managerial challenges.* *Managers must adapt their crisis management strategies accordingly.*
- *Sudden crises pose problems of pressure, stress and information management.* *An emergency plan can provide time for reorientation after the initial shock of a crisis. It also helps people cope with the rapid influx of information. Managers should ensure that financial responsibilities are not neglected in the early phases of a sudden crisis. There is a tendency to focus on technical and organisational problems to ensure the survival of the project.*
- *Creeping crises encourage procrastination by not appearing urgent.* *Managers should be vigilant to the tactical manipulations of interest groups which have time to formulate strategies to serve their financial interests. These tactics are a source of potential dispute and a dangerous distraction from the real problem, which may be technical or organizational rather than financial. If neglected, these problems will accumulate.*

MANAGING COMMUNICATIONS DURING A CRISIS

Chapter Five portrayed effective crisis managers as social architects who exhibit an intimate understanding of the forces that shape communication patterns among project members and of the impact that these have on crisis management outcomes. This is discussed in more detail here.

Selfishness

Chapter Three pointed to people's formal and informal responsibilities to protect specific project goals. A person's involvement in the management of a crisis should be determined by the extent to which the goals he or she protected were threatened. However, in the case studies, there was little evidence of this match between responsibility and crisis management involvement. The weak tended to be excluded from the process and the powerful dominated it. This caused a considerable degree of frustration and resentment for the suppressed.

Lessons:

- *Appreciate peoples' varied interests and the communication patterns that serve them.* *Monitor communication patterns that might damage the project.*
- *Assess a crisis' impact on project goals and ensure the appropriate specialists are involved to the appropriate extent.* *For example, if costs are primarily threatened, ensure the person responsible for budgetary control is in a position of power.*
- *The resolution of a crisis may demand trade-offs between different goals, bringing into conflict people who would otherwise work together.* *Managers should identify and focus energies on managing these tense interfaces since they are a source of potential disruption and inefficiency.*

The way the crisis was managed

The managerial style of the crisis manager appeared to have a significant influence on people's communication patterns during a crisis. People tend to become autocratic as a defence mechanism, often in an attempt to re-establish control. The outcome was, without exception, a loss of control.

> *Lesson:*
> - **Crisis management needs to be independent, fair, and open.** *Most project members including the crisis manager may have a vested interest in a particular outcome. It may be worthwhile considering third-party intervention or an external consultant to manage the process.*

Unexpected sub-crises

Unexpected sub-crises punctuated each crisis and significantly affected people's communication patterns. These sub-crises were the consequence of latent tensions from mismanaged crises and in this sense were all self-manufactured. Like the main crises they occurred within, their early warning signs were not detected and they had the potential to be destructive. However, they often had a positive influence, increasing the cohesion of the project team because they gave people a common focus and an opportunity to re-adjust their relationships.

> *Lessons:*
> - **Expect the unexpected during a crisis.** *Crises spurn sub-crises that can be creeping or sudden.*
> - **Occasionally it may be useful to stimulate a sub-crisis in order to increase group cohesion.** *This is like using an explosion to extinguish a fire and takes considerable courage.*

Personal relationships

Personal relationships also shaped peoples' communications during each crisis. These relationships appeared to be determined by experiences on previous projects, stereotype views of occupational groups, in-built traditional suspicions among certain occupational groups, and by experiences during the life of a project.

> *Lessons:*
> - **Under the pressures of a crisis, personal relationships are tested to the limit and any underlying tensions are exacerbated and exposed.** *Ideally, construct project teams using people who have positive relationships from past projects. Where this is not possible, use people with no previous relationship. Finally, monitor interpersonal relationships continually for new tensions which will indicate where problems are likely to arise during a crisis.*

Peer pressure

In many of the crises, peer pressure played a considerable part in shaping peoples' communication patterns. People of common interests tended to group into temporary coalitions that developed strong internal pressures to be loyal to fellow members and to conform to established patterns of behavior and group norms. The intention of this coercive behavior was to suppress a potential problem that could implicate a group in blame or to hide it from the view of more senior decisionmakers.

Lessons:

- **Look out for corporate bullying.** *During a crisis, some interest groups can become excessively powerful, forcing through solutions which are inequitable or not in a project's interests.*
- **Don't rely on people to communicate potential problems.** *The common interests that develop within projects can lead to protective behavior, particularly if someone is at fault. The absentee project manager who relies on the goodwill of individual consultants to inform him or her of problems is most vulnerable. To prevent the incubation of potential problems, managers must be intimately involved in a crisis, trusted and accessible to potential informers or whistle-blowers.*

CRISES AS PERIODS OF SOCIAL CHANGE

The previous discussion indicates that construction crises activate a process of social adjustment within its host organization. This adjustment process is the product of a constantly developing struggle for information that will enable people to force resource redistributions in their favor. During this struggle, individuals attempt to mold the structure of their personal social network and those of others to suit their own interest—interests that may not necessarily coincide with the client's. To this end, individuals may form interest groups that increase their power base. The structure of the communication network that eventually emerges is determined by the relative success of each individual or interest group in imposing his or her desired structure on others.

Lesson:

- **A crisis causes social chaos.** *During a crisis, "normal" loyalties and allegiances change, and it pays to be sensitive to the inevitable formation of temporary coalitions/factions. Their informal nature makes them invisible to the eye and difficult to detect. They are best identified by engendering an enduring sense of trust and confidence in one's judgment and leadership.*

The folly of imposed solutions

While individuals or groups may attempt to manipulate communication structure to serve their interests by, for example, marginalizing those who represent a threat, the case studies showed that the interests and power of those they sought to manipulate largely determined their success. While some people responded positively to the manipulation because the direction suited their own interests, others resisted vehemently. This process of reinforcement and resistance continued until the interests of the interacting parties were mutually served by the structure of their communication network or, alternatively, when one party was able to forcibly impose a communication structure on another. However, these forced social equilibriums turned out to be little more than a pleasant illusion because, they resulted in increased tension and frustration for the person who was suppressed. In the long-term, there was a limited amount of time over which this tension could be contained. Equilibrium was most rapidly and lastingly achieved when those involved had shared interests because they voluntarily arrived at the same solution.

Lessons:
- *Imposed solutions are not lasting solutions. The natural temptation during a crisis is to forcibly impose a solution. Although they may appear to work in the short-term, such solutions are temporary and create covert, latent tensions that compromise subsequent crisis management efforts or eventually manifest themselves in the form of a sub-crises. Ultimately, when a multitude of interests are involved, lasting solutions only come through a process that gives all parties a sense of ownership over the eventual solution. The extra time this takes is a good investment.*
- *Peoples' interests can be aligned by sharing risks, by helping people focus on common interests rather than their differences, and by encouraging people to put their preconceived solutions aside. Rather than traditionally searching for compromise solutions, help people work together to search for the solutions that may suit everyone's interests.*
- *Managers must control the balance of power during a crisis. They should be knowledgeable about a crisis and portray themselves as arbiters who are a source of independence, reliability, and fairness.*

EFFICIENT COMMUNICATION PATTERNS

This section discusses the development and efficiency of various communication patterns. With this knowledge, managers can predict, interpret, and thereby control information flow more effectively during a crisis.

Factionalism
The emergence of factions was discussed in the previous section, as was their tendency to develop around those with common interests. However, the range of

common interests that seemed to underpin the development of factions was varied, and factional membership did not always mirror the traditional interest groups we associate with construction projects. Such factions were useful because they broke down barriers to communication. However, unfortunately they were relatively rare because their formation often demanded considerable insight and courage to ignore traditional occupational stereotypes.

Most factions emerged because people wanted to increase their power base in negotiations, to manipulate information flow, or to defend personal interests by hiding a problem. These types of factions tended to develop more easily in environments where blame could be attributed to someone. On the other hand, factions also emerged for altruistic reasons, such as wanting to share information and to help solve a problem. They tended to develop when risks were shared and when people recognized their mutual interdependency.

Unfortunately, the negative reasons why factions developed outweighed the positive, which caused them to have a damaging affect by focussing communications into intense pockets, fragmenting the organizational structure, and leading to an overall breakdown of communications. There were some exceptions to this rule, and it was evident that factions were not a problem if their activities were monitored to ensure they had positive outcomes and if opportunities were provided for people to communicate with them. Indeed, effectively controlled factions could be useful to managers by simplifying the organizational structure of a project into fewer component groups with fewer interfaces.

Lessons:
- *A crisis has the potential to generate unpredictable and often surprising allegiances.* *Be careful, allegiances are not always predictable and the reasons behind their formation can be temporary and obscure.*
- *Managers should create conditions that promote the development of positive factions.* *This is done by encouraging interdisciplinary teams, by sharing risks, by being non-recriminatory and by emphasizing people's interdependence.*
- *Managers should be vigilant to the formation of negative factions and disband them.* *Negative factions can be recognized by their secretive nature and by their suspicion of, and resistance to, outsiders. Negative factions often include people who are afraid of being blamed for a crisis.*

Centralization

In many instances within the case studies, people sought to occupy central positions within the communication networks. They did so for a variety of reasons and in a number of different ways.

Regaining control

Some people sought central positions to gain control. Often, they did this by attempting to dominate information supply and/or demand. Another tactic to gain control was to seek gate-keeping positions within the communication network that would permit them to filter and manipulate information to their own ends.

Avoiding blame

The high stakes associated with a crisis exacerbate the sense of fear about being implicated in blame. One way to avoid blame was to reduce one's directness of communications with those who were implicated in the problem. By working through intermediaries, a person could distance him or herself from a problem and be seen as merely acting on the advice of others.

Avoiding bias

Many vested interests are at stake during a crisis and information blockages and distortions can arise as people in gate-keeping positions attempt to block its transfer between certain people. One way in which people attempted to overcome this barrier was to find alternative routes around these potential sources of bias.

Gaining power

The differences of interests that emerge during a crisis produce different interpretations of what a crisis is about and who is responsible for it. This often led to a process of negotiation during which information was an important source of power. In full knowledge of this, some people sought to ensure that their information networks were unique and they cloaked them in secrecy. On the other hand, opponents sought to infiltrate these communication networks in an attempt to reveal the basis of their arguments and ultimately undermine their case.

What was the most efficient pattern of information flow?

In terms of efficiency, the best pattern was one that had a high level of centrality for information supply but a low level of centrality in information receipt—that is, where there was a widespread supply of information from a restricted source. This resulted in a strong sense of leadership and a minimum level of misunderstanding because everyone was operating on the basis of similar information. Advantage also seemed to arise from "denser" communication networks where everyone implicated in a crisis could communicate directly with each other without having to go through intermediaries who could be biased. In dense networks, people are tightly knit and close to each other. This provides flexibility, enabling problems, ideas, and solutions to spread rapidly with little distortion.

Particularly high levels of efficiency seemed to arise when the communication network was tightly knit around one person because this ensured a high degree of consistency in communications. Another advantage of high density was that it minimized the possibility of gatekeepers emerging by providing a variety of alternative routes through which information could flow. The danger of gatekeepers was that they focussed information through restricted channels and thereby increased the possibilities for information overload and bottlenecks. They also provided opportunities for people to manipulate information in their own favor. The influence of gatekeepers was not always negative, being determined by the personal characteristics of people occupying those key positions. When they were manipulative, negative, and unable to cope with the information demands on them, the effect was damaging; when they were experienced, capable, and conscientious, the effect was positive.

Finally, a high level of similarity (equivalence) between people's communication networks equalized information differences and ensured that everyone was working on the basis of the same information. In contrast, when the level of similarity was low, the organization became disjointed and characterized by people pulling in different directions.

> ### Lessons:
> - **The patterns of information flow during a crisis are an important determinant of crisis management efficiency.** *As far as possible, managers should build communication structures with few negative factions, many positive factions, a restricted source and widespread supply of information, a high level of density (directness in communications), a single point around which communications are tightly knit, few gate-keeping positions, and a high level of similarity in contacts at individual level.*
> - **Since communication patterns determine people's power-base, they become the subject of considerable manipulation.** *This may cause information filtering, distortion, or restriction in certain directions, harming crisis management efforts. Generally, people seek positions of high centrality and try to differentiate their communication network from those of others. Managers should ensure that the people with the right motives are in the central positions and that people's communication networks are well-integrated.*
> - *A warning: While an efficient communication structure can help ensure open and clear information flow, it cannot guarantee crisis management success. This is also determined by the quality of information that is transferred within it. This, in turn, is largely influenced by the motives and competencies of its members. No system, however well-designed, can compensate for the effects of people who are determined to sabotage or who are unable to cope with the pressures of a crisis. In this sense, attention to positive relationships and to the motives and personal qualities of people affected by a crisis represent the foundations of efficient crisis management.*

CRISIS BEHAVIOR

All of the crises demanded a major investment of energy, time, and resources beyond that envisaged at the start of a project. However, the efficiency with which this extra energy was expended varied, each crisis being punctuated by change points that separated distinct phases of behavior and forward momentum.

Behavior during creeping and sudden crises

The case studies included examples of creeping and sudden crises. When their behavioral characteristics are compared, some interesting patterns emerge. For example, in contrast to sudden crises, creeping crises are more emotionally charged, divisive, and problematic for managers during their early stages—precisely the opposite of what would be expected. The explanation may be that the shock of a sudden crisis forces people to temporarily abandon their differences in an attempt to secure the immediate survival of the project. During a creeping crisis, people have more time to selfishly pursue their own interests, resulting in neglect of important problems that gradually accumulate in proportion until a second, but sudden sub-crisis forces people to put their differences aside.

Lessons:
- ***Surprisingly, people's behavior during a creeping crisis is more unstable and dangerous than during a sudden crisis.*** *Sudden crises create a sense of urgency and early forward momentum that can continue, if managed effectively. This is achieved by identifying, reinforcing, and supporting early successes and the positive working relationships that emerge. In contrast, creeping crises create little sense of urgency, allowing people to pursue personal interests. Managers must be sensitive to the covert tensions that often represent the early warning signs of a creeping crisis. One way of overcoming the lack of urgency and division that often characterize creeping crises is to artificially generate a sudden crisis. If managed effectively, the positives that emerge can be used to diffuse the differences that were fuelling it.*

The unpredictability of crisis behavior

People's behavior during a crisis is unpredictable. For example, the earthwork support crisis (case study one) began with a phase of negative momentum, indecision, uncertainty, formality, rigidity, defensiveness, and escalating conflict. It proceeded to a second phase characterized by forward momentum, mutual sensitivity, open communication, attentiveness to the problem, collective responsibility, low uncertainty, and low emotions. The crisis then fell back into a third phase of negative momentum, uncertainty, confusion, and heightened emotions. Finally, it returned to a phase of forward momentum, cooperation, decisiveness, negotiation, compromise, collective responsibility, and low emotions.

While the paviour crisis (case study four) followed a similar pattern, the retaining wall crisis (case study three) induced an opposite pattern of behavior, beginning with a period of strong forward momentum, widespread commitment to resolving the problem, open discussion, collective responsibility, and relatively low emotions. It then proceeded to a phase of zero momentum, increasing uncertainty, indecisiveness, reduced attention to information supply, growing frustration, and anxiety. Finally, it moved into a deeper phase of negative momentum, inflexibility, confrontation, and heightened emotions. In complete contrast to any other crisis, the elevator crisis (case study two) was characterized by a consistent sense of forward momentum and no significant changes in behavior.

> *Lessons:*
> * ***One cannot generalize about a common pattern of behavior emerging in response to a construction crisis.*** *Beware of universal models of crisis management. There are no quick fix solutions to construction crises and the unpredictable patterns of behavior that emerge, demand a more intelligent, thoughtful, and responsive approach to crisis management founded on a clear understanding of what motivates people to behave in certain ways.*

Predicting changes in people's behavior

If the crisis management process is likely to move through self-perpetuating periods of forward and backward momentum, the crisis manager is challenged to get it into an accelerating mode and to keep it there. This requires knowledge of the conditions that induce positive behavior, an understanding of what causes behavioral changes, and sensitivity to the advance warning signs of change.

Major change events

Every change point in every crisis coincided with a major event such as the surprise serving of a claim or the sudden intervention of a senior manager or of the client. While the nature of the "change events" varied, they were all manufactured by those involved in the crisis management process—sometimes deliberately but at other times, unintentionally. Furthermore, while each of these change events seemed to be sudden, numerous signs of impending change preceded them. For instance, in one crisis, where the involvement of the contractor's regional surveyor induced a sudden change in behavior, there was a detectable and steady increase in tension before the event. This manifested itself in the contractor's constant warnings and threats of delay. Similar warning signs preceded all the major change events.

Despite the existence of early warning signs of behavioral change, people showed insensitivity to them, sometimes voluntarily and at other times, involuntarily. For example, in several cases, people deliberately ignored their opponents' obvious frustrations and their warnings that they were going to escalate a dispute. However, there were also instances when people (normally senior decisionmakers) were insulated from these warnings by those who stood to lose from exposure to blame. The most disturbing fact to emerge was that as time went by, insensitivity and resistance to change grew because accumulating losses magnified the negative implications of accepting change. At the same time, the parties who wanted change became increasingly forceful in pursuing it. This created further resistance to change, and so on, until the project team became drawn into a spiral of insensitivity, acrimony, and escalation. That is, when a project entered a negative phase of behavior, it deepened rapidly and became increasingly difficult to stop.

More disturbing was the evidence that insensitivity to change is as much a problem in positive periods as it is in negative periods. For example, in several instances, the euphoria of progress and of strengthened project relationships blinded people to the real tensions that existed within their projects. Thus, in contrast to negative phases of behavior, which are self-reinforcing and robust, positive phases are fragile and easily destroyed.

While behavioral instability seems inevitable, one crisis (case study two) was unique in its absence of behavioral change points. In this crisis, the architect's sensitivity and responsiveness to the contractor's needs and emotions meant that he would always take action to supply information in time to counteract any escalation of the crisis. Similarly, the contractor's site manager's determination to make the project a success meant he did not seek to exploit the architect's misfortunes and was willing to demonstrate flexibility in helping resolve the crisis. In effect, the contractor performed an important compensating function that helped nurture a relationship of mutual trust and sensitivity that in the long run, reinforced the stability of the crisis management process.

Lessons:
- *If everything is going well you have probably overlooked something.*
- *The most efficient crisis management processes are the most behaviorally stable. Behavioral instability is not inevitable during a crisis. The challenge is to encourage and then maintain positive thinking. The most effective way to do this is to encourage mutual sensitivity to the needs of others, open communication, and collective responsibility for dealing with crises.*
- *The importance of early intervention in crises cannot be over-stated. Negativity is self-perpetuating and over time, is increasingly difficult to break.*
- *Crises have in-built defense mechanisms because they create conditions that produce insensitivity to the advance warning signs of escalation.*
- *Crisis managers can stabilize behavior by being constantly vigilant to the signs of changing interpersonal relationships.*
- *Try to see crises as opportunities to increase team cohesion rather than as threats that can destroy it. Most important, encourage others to see it in the same way.*

Explaining behavior during a crisis

To control people's behavior, it is important to understand what causes it. People's behavior during a crisis appears to be motivated by a number of factors.

Stereotypes

The case studies indicate that people enter projects with pre-conceived stereotype images of other occupational groups that are largely shaped by their experiences. These stereotypes influenced people's initial behavior in response to a crisis, particularly early in a project's life cycle. For example, in one crisis, tensions between the consultants and contractor were rooted in an acrimonious dispute stemming from a previous project; in another crisis, the architect's early behavior toward the contractor was partly determined by a "bad experience" with a contractor on a previous project. While we would expect the behavioral influence of stereotypes to be at their strongest during the early phases of a project, the case studies indicate that they are persistent, enduring, and resistant to change.

Lessons:
- ***Although project relationships may seem tranquil and amicable on the surface, there may be underlying tensions within the project team.*** *The high resourcing demands, ambiguities, and pressures induced by a crisis create ideal conditions for these tensions to surface and grow.*
- ***Stereotypes are dangerous because by definition they are arbitrary, overly simplistic, and negative.*** *Stereotypes have minimal influence in teams whose members have had positive experiences with each other and been kept together for the life of a project and between different projects. Success breeds success and familiarity erodes the ignorance that is the basis of stereotypes.*

Financial responsibility

People's behavior was also shaped by the degree of uncertainty surrounding financial responsibility for a crisis. Uncertainty primarily arose from ambiguities in contractual documentation that were exploited by those who were most likely to be held responsible for a crisis. High levels of uncertainty surrounding contractual responsibilities enabled potential "losers" to redefine events in their favor, creating further uncertainty and prolonging the crisis management process. In contrast, in an environment of financial certainty, people's behavior was characterized by open discussion, clarity, and a sense of forward momentum.

> **Lessons:**
> - **Share project risks equally whenever possible.** *With equally shared risks, the issue of financial responsibility is irrelevant and the focus becomes the collective minimization of mutual losses rather than the transfer of them.*
> - **When risks cannot be shared, patterns of risk distribution should be clarified and a mutual understanding of them developed.** *It is an illusion that using a standard contract automatically increases mutual understanding and clarity. Risks are best clarified by simplifying contracts, minimizing their number, and insisting that people who are privy to them, discuss them.*

Fear

Not sharing project risks caused people to focus on their differences rather than on their similarities; to create an atmosphere of fear, recrimination, and blame; and to shut down communication channels precisely at the time when increased communication was needed. However, in one case study, the project team managed to overcome built-in contractual divisions by focusing on ways to minimize the costs of a crisis rather than on resolving responsibilities to pay for it. The result was cost implications that were not significant enough to argue about.

> **Lessons:**
> - **Focus on how to minimize costs rather than on who will pay for them.**
> - **Focus on people's similarities and mutual interdependencies rather than on their differences and independencies.**
> - **Avoid a recriminatory response to a crisis.** *Depersonalize investigations, refrain from allocating fault or blame, focus on solutions rather than causes, and create a supportive rather than penal environment.*

Organizational policies

People's behavior was also guided by the policies of their employing organizations. For example, in one case study, the contractor encouraged the pursuit of claims, which likely influenced the behavior of its employees. In the same project, the consultants' defensive response was probably guided by the client's strict policy on budgetary control. In contrast, in the most successfully resolved crisis, the contractor's overriding company policy was to be as cooperative as possible in order to make a favorable impression on a new client. The resultant tolerance and accommodation this policy encouraged within its workforce contributed toward the effective resolution of the crisis.

> **Lessons:**
> - **When constructing a project team, give as much attention to the cultures of the companies as to the personalities and attributes of individuals.** *Beware of aggressive, selfish organizations. This includes the client.*
> - **It is difficult to maintain a balance, between controlling project goals and providing the flexibility required to avoid conflict.** *Inflexibility engenders defensiveness and rigidity in the bargaining process, which leads to escalation. Goal flexibility is not about publicizing acceptable tolerances on projects goals because this will induce a lack of diligence. Flexibility means being prepared to work within contingency allowances and in this respect, it is essential that organizations are permitted to price for their risks.*

THE TWO IRONIES OF CONSTRUCTION CRISIS MANAGEMENT

Crises have built-in defence mechanisms that cause people to behave in ways opposite of what is necessary for effective resolution. In particular, it is ironic that at a time when collective responsibility and teamwork are important, conflict is more likely, and at a time when effective communication is important, it is less likely.

Irony #1: When a sense of collective responsibility and teamwork are more important, they are less likely to materialize.

In explaining this irony, we return to the issue of risk distribution practices and the tendency for crises to demand a significant injection of extra resources into a project.

In essence, the main problem with separate risk distribution is the emergence of distinct winners and losers in the resource redistributions that inevitably follow a crisis. This generates a lack of collective responsibility for resolving the crisis. Problems arise as different parties turn to their formal contracts to clarify patterns of responsibility, only to discover differences in interpretation and misunderstandings. If it were not for the crisis, these differences would remain irrelevant, but their effect is to create an air of uncertainty and to detract attention from the resolution of a crisis. Contractual ambiguities also provide opportunities for parties to employ a range of informal bargaining tactics to force resource redistributions in their favor. In this sense, formal contracts only partially determine patterns of responsibility during a crisis; a party's tactical prowess and bargaining power are also strong determining factors.

Tactical miscalculations and unintentional escalations

The previous section illustrates how crises, when coupled with separate risk distribution, can lead to fundamental changes in the nature of interpersonal relationships within a project team by exposing conflicts of interests which would otherwise remain hidden. Any disagreements that emerged during a crisis were made more contentious by the low margins under which parties were employed and the high stakes that, by definition, each crisis created. Collectively, these conditions magnified any conflicts of interest between project participants, making them more sensitive to money-making opportunities and more resistant to potential losses. In this situation, parties were prepared to use a variety of bargaining tactics to force resource redistributions in their favor. For example, in one case study, where a contractor served a claim for extra monies, consultants responded by generating outlandish counter-arguments to call their bluff and test their resolve in pursuing it. There were also numerous examples of parties attempting to take advantage of any ambiguities they could find in contractual clauses or in the nature and causes of the crisis itself. The popularity of this tactic seemed related to its low risk of escalation. However, another common tactic with a higher risk of escalation was to ignore a counterpart. Such a tactic always produced warnings and threats and eventually, deliberate acts of escalation. Indeed, in all but one of the crises, the combination of tactics employed by the various stakeholders had an escalating effect. Assuming that no party wanted to deliberately precipitate a full-blown conflict—and there was no evidence to the contrary—the observed bargaining processes can be seen as a catalog of tactical errors.

> ***Lessons:***
> - ***Crisis managers must manage.*** *This is particularly important during a crisis, when there is a natural tendency toward escalation if there is no intervention.*
> - ***Crises highlight and exaggerate any weaknesses, misunderstandings, and divisions within a project team.*** *At a time when collective responsibility is more important, conflict is more likely.*
> - ***The more serious a crisis in terms of resourcing implications, and the lower the margins of implicated organizations, the more passionately they will defend their interests and the more likely it is that conflict will occur.***
> - ***Contracts afford little protection against parties who are determined to avoid responsibility.*** *Tactically, parties wield considerable illegitimate informal power to force resource redistributions in their favor, despite what the contract says. Party's tactical astuteness rather than contractual obligations are the main shaper of resource distributions during a crisis.*
> - ***When parties abuse their illegitimate power, weaker parties, who may be in the right, suffer unduly.*** *This causes malevolence, frustration, and the potential for conflict. Try to ensure fairness in the bargaining process.*
> - ***The process of crisis management is fraught with the danger of accidental escalation resulting from miscalculated tactics of differing interest groups.*** *Be vigilant to potentially dangerous tactical combinations.*

Dangerous tactical combinations

Managers should be aware of and vigilant to tactical combinations that can produce accidental escalations of disputes. With knowledge about the motives that drive such tactics, managers are better equipped to understand a dispute, to predict its course, and to identify workable solutions with a low risk of escalation. To identify potentially dangerous tactical combinations it is useful to analyze the tactical patterns that characterized each of the case studies. These are depicted in Figures 10-1, 10-2, 10-3 and 10-4. The bargaining codes refer to those depicted in Table 5-1.

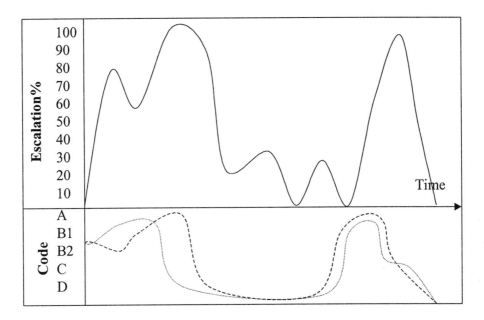

Figure 10-1 Tactical patterns across case study one compared to levels of escalation.

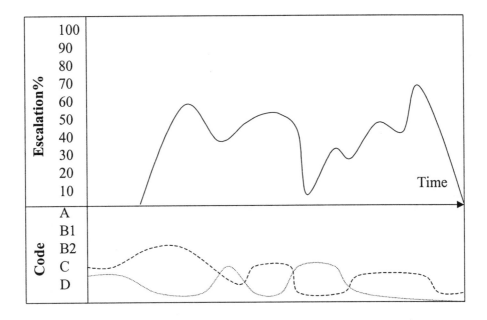

Figure 10-2 Tactical patterns across case study two compared to levels of escalation.

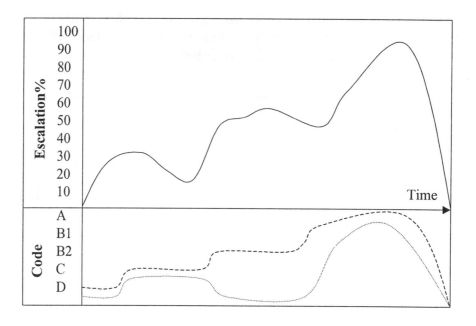

Figure 10-3 Tactical patterns across case study three compared to levels of escalation.

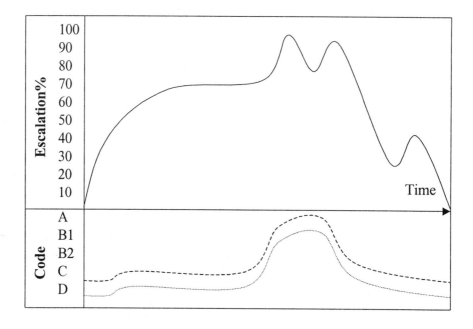

Figure 10-4 Tactical patterns across case study four compared to levels of escalation.

This analysis of the tactical patterns suggests that opposing parties tend to follow parallel routes. That is, one party's adoption of aggressive tactics is likely to eventually produce a similar response from an opponent, leading to an escalation of the situation. The potential for this tandem escalation seems to be related to the equitable balance of power that exists in construction projects. While the legitimate power that derives from contractual conditions may enable one party to forcibly suppress an opponent's case, parties to a construction project are also able to wield a considerable amount of illegitimate power, which most seem prepared to use. To prevent escalation, one party needs to break the mold and show a willingness for conciliation, compromise, and collaboration. Such a commitment is likely to induce a similar response from an opponent. However, a move to a conciliatory bargaining code must be genuine and be perceived as such by an opponent. If it is not, it will not alleviate the problem.

Lessons:
- *Contracts do not resolve crises. People do. Contracts are no substitute for good management.*
- *Effective communication between opposing parties is critical to avoid accidental escalations. Most disputes are based on simple misunderstandings.*
- *Coercive power tactics cannot bring about a successful long-term resolution of a dispute, nor can those based on compromises. By definition, compromises result in sub-optimal solutions and do not fully resolve underlying differences and tensions. Coercion and compromise merely create the illusion of a solution.*
- *Collaboration is the only way to fully resolve the tensions generated by a crisis. This is because it seeks optimal solutions that benefit all interest groups. However, this process is time-consuming and it demands courage to let go of preconceived ideas and creativity to generate imaginative alternatives.*

Irony #2: At a time when effective communication is important, it is less likely.

Good communication is essential to efficient crisis management because information is the antidote to the uncertainty that crises generate. Unfortunately, at a time when effective communication is of particular importance, crises tend to create conditions in which it is less likely to occur.

Information as a source of power

Under the exaggerated conflicts of interest that emerged during the case study crises, information became an increasingly valuable commodity. This was because it represented an important source of power in negotiations and, consequently, it was more closely guarded. Indeed, the inclination to withhold information was magnified by the tendency of people with similar interests to collect into groups to share information for their common good. Such groups developed a sense of secrecy and pressured their members to put the group's interests before their own.

The volume of information generated

Another barrier to effective communication was the shear volume of information generated during a crisis, coupled with a contraction of responsibility for decisionmaking. This was a particular problem in the sudden crises, when "hot points" emerged within the organizational structure, creating bottlenecks and information overload. This caused information demand to run ahead of its supply. Unable to cope, the people occupying those positions fell into an increasingly reactive style of management. Some people attempted to cope by insisting that formal and standardized procedures be adhered to while others coped by becoming increasingly informal. However, both coping strategies deepened the problems they were designed to resolve. Formality slowed down information supply and caused frustration, and informality lead to breakdowns in communications and misunderstandings.

Lessons:
- ***People naturally withhold information during a crisis because of exaggerated conflicts of interests.*** *Encourage communication during a crisis by emphasizing common interests and by having regular "communication meetings" when information demand and supply is the only agenda item. Use these multi-disciplinary meetings to generate and update information plans and to clarify information interdependencies and responsibilities for information supply.*
- ***Information quality is as important as quantity.*** *Alleviate the time pressures that cause people to make mistakes. Employ competent people.*
- ***"Hot points" emerge during a sudden crisis.*** *Identify the people who might occupy these points and support them. These people have the most critical positions in relation to information flow but ironically they are the most vulnerable to the stresses that can damage it.*
- ***It is as inappropriate to deal with a crisis in a routine, formalized manner, as it is to respond in a completely informal manner.*** *The emphasis should be on a balance of flexibility and control. Unfortunately, during a crisis, this type of finely balanced response is less likely. Instead, project participants exhibit extremes of formal or informal behavior that lead the project into a downward spiral of poor communication, tension, anxiety, and stress, which becomes increasingly difficult to break.*

THE DYNAMICS OF CRISIS MANAGEMENT

To conclude this chapter, we turn back to the model of crisis management presented in Figure 3-1 and to the numerous inefficiencies in detection, diagnosis, decisionmaking, and implementation that created inertia within the crisis management process in each case study.

Monitoring

The case studies provided numerous instances of insensitivity to the early warning signs of potential crises. This insensitivity was caused by ineptitude, changes in project teams during a project, the timing of events in dormant periods, defensiveness, and fear of blame. Indeed, the blindness to potential problems was so extreme in some cases that the project's environment was more sensitive to problems within the project than the project team themselves. This highlights the importance of including all stakeholders in the management of a project.

In addition to monitoring inefficiencies there were also problems in the transition between the phases of monitoring and diagnosis. That is, knowledge of potential problems was often withheld from decisionmakers—sometimes deliberately, due to conflicts of interests and a lack of trust between monitors and comparators.

Diagnosis

Of all the phases in the crisis management process, diagnosis emerged as the most problematic. It was here that the financial responsibility for a crisis was determined and that potential winners and losers fought to safeguard their interests. In many instances, those in positions of contractual power dominated the process of diagnosis and manipulated it to protect their interests by blocking the exposure of potential blame to higher-level decisionmakers.

Decision making and implementation

There were also inefficiencies in the decisionmaking phase of the crisis management process, largely because, under the pressures of a crisis, people tended to rush their decisions and base them on inadequate or inaccurate information. In sudden crises, the focus was on organizational issues at the expense of financial issues, while in creeping crises, the opposite was the case. Consequently, financial issues emerged as a problem during the implementation phases of sudden crises and organizational issues emerged as problematic during the implementation stages of creeping crises.

Learning and recovery

The landscape of interpersonal relationships changed considerably during a crisis—in most instances, for the worst rather than for the better. However, there was no evidence of any attention being given to the process of learning and recovery. Even in the most successfully resolved crisis, people's preference was to put the sager behind them and to "gloss-over" over the conflicts of interest that had emerged within the project teams.

Repeated cycles of the crisis management process

Paradoxically, while the managerial inefficiencies, introduced inertia into the crisis management process, the effect was to force it through repeated cycles. For example, the earthwork support and lift crises went through two cycles, the paviour crisis went through three cycles, and the retaining wall crisis went through four cycles. In each case, repeated cycles of the crisis management process were necessary because inefficiencies created tensions among people who felt that their needs had not been satisfied. Each subsequent cycle was the mechanism by which unsatisfied and oppressed parties attempted to alleviate this tension. The more tension, the more cycles were needed to alleviate it. For example, in one crisis, an architect's reluctance to countenance a contractor's claim for extra monies led to delays on site which, in turn, led to another claim for an extension of time, which initiated another cycle of the crisis management process. The crisis management process will continue to go through repeated cycles until a point of equilibrium is reached where all the latent tensions associated with previously unsatisfied needs are dissipated.

Lessons:
- *Provide incentives for people to monitor and communicate potential problems regardless of who bears their risk. This includes all stakeholders (external and internal to the project team). Do not ignore the opinions of external stakeholders.*
- *Clarify mutual interests and interdependencies in resolving potential problems rapidly.*
- *Keep project teams as consistent as possible for the duration of a project.*
- *Beware of dormant periods during a project when people's attentions are focused elsewhere.*
- *Encourage and facilitate rapid, firm, and clearly communicated decisions.*
- *Avoid rushed, ill-considered decisions, and ensure they are based on complete and balanced information relating to technical, financial, and organizational issues from a broad range of perspectives. People tend to panic during the early stages of a crisis and prioritize issues. It is more efficient to take time to get it right the first time than to spend time coping with the aftermath of a poor decision.*
- *Ensure that every stakeholder in a crisis feels his or her interests have been fairly represented in the crisis management process. Unresolved interests will initiate repeated cycles of the crisis management process. The most efficient crisis management processes move through one cycle only.*
- *Give attention to learning and recovery. De-briefing is an important aspect of crisis management. This involves discovering and attending to damaged relationships and initiating independently managed post-mortems on the crisis management process.*

CONCLUSION

This chapter draws similarities and contrasts among the case studies described in the four previous chapters. The goal was to identify important practical lessons to help managers resolve construction crises. The discussions highlight the importance of planning for crises and the dangers of complacency in managing them. All crises appear to have built-in defence mechanisms that create conditions that make their management more difficult. In particular, they all have the potential to bring otherwise hidden tensions to the surface, to damage communications and interpersonal relationships, to cause people to behave selfishly and manipulatively, to generate further problems, and to gather their own destructive momentum. While the lessons that have been highlighted in this chapter can help mitigate these problems, their main purpose is to help managers realize their most important lesson: that a crisis can be turned to advantage if it is managed effectively.

Chapter 11

Conclusion – Creating an Optimistic Organization

While crises undoubtedly hold dangers for managers, they also present unique opportunities for improvement. This chapter summarizes the organizational attributes that can unlock this productive potential. This summary has both cultural and pragmatic elements. The pragmatic element takes the form of guidance for managers who are faced with a crisis; the cultural element is more generic and supportive. The chapter begins by focusing on cultural issues because without a positive culture, nothing pragmatic is realistically achievable.

OPTIMISM OR PESSIMISM – A STARK CHOICE

This book began with a sense of optimism for the future. It ends with the realization that managers can control their destiny by seeing uncertainties and complexities as threats or as opportunities. While this stark statement seems simplistic, it does not mean that managers should ignore the undeniable threats that crises pose. It does illustrate that managers can approach crises with a positive or negative mind-set, the former being differentiated from the latter by a determination to turn them to advantage by focusing on their opportunities. Most managerial practices in the construction industry are pessimistically guided by the principle of mitigation rather than optimisation. The case studies have demonstrated that once a project commences, traditional contracts, organizational practices, and cultures, lock people into a set of predetermined performance standards with little incentive to improve on them. They also create a penal culture of suspicion, fear, and mistrust that stifles the openness and freedom which defines the innovative, courageous, and altruistic culture of an optimistic organization. In the case studies, the greatest success seemed to be enjoyed by those people who were prepared to fight the system.

THE OPTIMISTIC ORGANIZATION

The discussions throughout this book have highlighted the principles that underpin an optimistic organization. These are discussed below.

Optimistic people
The defining characteristic of optimistic organizations is their determination and ability to see opportunities for performance improvement and their courage to take advantage of these opportunities. This demands a positive mind-set that can

only be achieved by employing people with a sense of optimism, courage, energy, and determination. However, every individual on a project is employed by and to some extent controlled by an organization, and if relationships at the organizational level are not positive, then any sense of individual optimism can rapidly disappear. This is a problem for construction project managers who are often faced with the challenge of motivating individuals who are working for demotivated and even resentful directors. In this sense, generating optimism will depend on changing the construction industry's traditional employment practices, contracts and organizational structures. These practices developed within a relatively stable but highly confrontational environment and were designed to make it easy to blame someone when something went wrong. Consequently, the principles that underpinned them were centralization, uniformity, hierarchy, compartmentalization, prescription, and rigidity. However, in today's opportunistic environment, such principles are obsolete since their effect is to fuel a self-perpetuating cycle of pessimism that suppresses the individuality, innovation, creativity, and courage that allow organizations to take full advantage of the future. A more optimistic construction industry will depend on new models of operation based on the opposing principles of flexibility, openness, equality, interdependence, collective responsibility, trust, sharing, and understanding. Making this transition can liberate the industry's workforce from the atmosphere of fear that has limited its potential productivity.

Effective communication

Clearly, good people alone cannot guarantee success. Good people need the support of an effective communication system that facilitates the sense of trust and openness that the industry needs. Information shortages are the main cause of division and mistrust during a crisis, and a well-designed communication system should prevent bottlenecks, disperse information rapidly and evenly, and minimize the potential for manipulation.

In addition to the cultural and organizational barriers to communication in the construction industry, there is the growing obsession with efficiency and leanness. The industry must be wary of this trend, since it will eradicate the redundancy needed to make communication systems reliable. Furthermore, the industry must change its restrictive contractual practices to enable people to legitimately opt out of formal procedures if they are restrictive and counter-productive during a crisis.

Preparedness

A sense of optimism depends to a large extent on an organization's ability to look for things that can go wrong. Organizations that know the future are better equipped to exploit the opportunities it may hold. It is worth reiterating that the world's most successful organizations are those that take time to compile prioritized crisis plans that demonstrate an in-depth understanding of their past and future vulnerabilities. However, confronting your organization's vulnerabilities means more than merely understanding the risks it faces. It is also about understanding its crisis capabilities, which means that the process of crisis planning involves realizing your organization's strengths as well as its weaknesses. This allows an organization to plan to maximize its strengths and minimize its weaknesses.

Another important aspect of optimistic organizations is that they are not over-optimistic. Optimistic organizations have a sense of realism in their plans and appreciate that the future cannot be planned accurately. Consequently, they are vigilant to continually changing risks and opportunities and adapt their plans accordingly.

Learning

A sense of optimism is founded on the belief that one can improve things by learning from past mistakes. Although this is often painful, many of the most profound sources of new knowledge are in the lessons to be learned in the aftermath of a crisis. Thus optimistic organizations have self-critical procedures in place to retrospectively dissect their crisis responses and to implement changes that can eradicate any future weaknesses. In contrast, pessimistic organizations prefer to put the past behind them and in doing so, fail to exploit what might be uncomfortable but valuable memories.

Securing early intervention

Crises tend to cause division and conflict, and thereby rapidly destroy any sense of optimism that may exist in an organization. For this reason, an important aspect of opportunism is early intervention, which depends on identifying potential problems quickly and nipping them in the bud. This is best achieved by eliminating fear, sharing risks, and instilling a sense of duty to notify potential problems, however small they may be. In essence, optimistic organizations act as confident, unified, self-supporting entities that are sensitive to all potential problems.

Creating a supportive and stable environment

The ability to achieve an opportunistic mind-set depends on the mental well-being and health of an organization's members. They need to work in an environment that is effective, controlled, stimulating, positive, and supportive of new ideas. Opportunistic organizations provide a sensitive, caring, and rewarding environment that minimizes uncertainty, prevents isolation, and minimizes frustration. Such an environment is possible by keeping project teams consistent, by minimizing the number of interfaces within a project team, and by reducing indecision and unnecessary change.

Creating a collaborative environment

Some degree of conflict accompanies a crisis, focussing people's minds on their differences rather than on their similarities. We have seen how conflict can be resolved positively if people move away from the compromising approach that is perceived as the most positive method of dispute resolution in the construction industry. The time and budgetary pressures that characterize construction projects drive the tendency to compromise; ironically, the process rarely alleviates those pressures. In contrast, collaboration provides the basis for exploring innovative and mutually beneficial ideas and solutions that can improve life for everyone.

The Pragmatics of Optimism

Managers, who put in place the preconditions for an optimistic culture, have the foundations for effective crisis management. This is management that turns crises to advantage and this book has offered practical guidance to help achieve this. Before the points of good practice are summarized, it is important to appreciate that this book offers no quick-fix solution. There is no substitute for thoughtfulness, sensitivity and responsiveness in dealing with the unique chain of events which precipitate every crisis. However, in general, to turn a crisis to your advantage:

- Do not ignore it.
- Do not procrastinate.
- Do not rely on contracts as a substitute for good management.
- Communicate the gravity of the situation. Everyone must know there is a crisis.
- Instill a sense of urgency if it is absent.
- Give priority to the crisis and resource it.
- Treat every crisis differently.
- Implement generic crisis management plans to buy some time.
- Identify an independent crisis manager to lead the crisis response as soon as possible. This may need to be an outsider.
- Define the crisis as soon as possible. Identify and involve all stakeholders because each will have a different perspective.
- Create a crisis management team. Involve necessary specialists, high-level managers, and main stakeholders who can contribute to a solution.
- Communicate the identity of the crisis management team. Define their powers and responsibilities.
- Loosen procedures. Do not be prescriptive about the way the crisis management team operates. Let them find their *modus operandi*.
- Define crisis management goals. Set targets to improve performance.
- Expect and demand results. Monitor and measure progress toward a solution.
- Be prepared to trade-off project goals.

- Present the crisis as a challenging opportunity by emphasizing the positives that can emerge.
- Care for every stakeholder. Independence is critical in crisis management.
- Having defined the crisis, implement detailed plans if they exist. They will not be a perfect fit and a detailed response will need formulating.
- Indicate clear concern for stakeholders outside the crisis management team.
- Keep everyone—insiders and outsiders—informed at all times.
- Do not speculate about responsibility or blame. Move on and solve the problem.
- Focus on minimizing costs rather than on who will pay for them.
- Encourage open communications. Information eliminates potential misunderstandings and conflicts. Continuous information is better than bursts.
- Be sensitive to people or groups who are able to control information flow. They will do so to serve their own interests.
- Encourage collective responsibility. Everyone's future is connected.
- Focus on relationships rather than individuals.
- De-personalize a crisis.
- Avoid blaming individuals or organizations. Look to the future not to the past.
- Do not talk about losers and winners. Everyone must feel that they can win.
- Identify people who see themselves as potential winners and losers. These represent the sources of tension and potential conflict.
- Focus on potential losers since they are the primary source of resistance to solutions.
- Monitor tactics between potential winners and losers. Encourage collaboration and be vigilant to the signs of deteriorating relationships. Beware of aggressive, selfish organizations—including the client.
- Look for and stop corporate bullying between project members.
- Do not impose solutions, other than as a last resort.
- Avoid quick-fix solutions. Rushed solutions return as problems.
- Look out for satellite problems that could become new crises. Crises create more crises.
- Do not neglect other aspects of an organization's activities.
- Communicate progress regularly. Focus on positives, not negatives.
- Encourage creativity and radicalism. Crises are extreme events that often demand extreme solutions.
- Stay cool. The symptoms of panic fuel crises.
- Identify "hot points" where stress can develop. Support the people that occupy them.
- Provide counselling/medical support where necessary. Show you care.
- In the aftermath of a crisis, provide positive feedback, conduct post-mortems, learn lessons, re-evaluate goals, re-focus the organization, repair damaged relationships, manage investigations sensitively, and move forward with a greater knowledge of your organization. You'll be stronger for the experience.

CONCLUSION

This book ends on a positive note. While it began with an enduring image of an industry that is struggling to cope with uncertainty, it ends with some insight into how the industry's apparently hostile environment can be turned to a manager's advantage. This insight has provided practical guidance of how to cope with crises in a positive manner. It has also provided a more holistic view of the supporting culture needed to underpin such efforts. The task of creating this optimistic culture is one of the most exciting challenges construction managers face.

REFERENCES

Abrahamson, M. W. 1984. Risk management. *International Construction Law Review* 1(3): 241-64.

Anderson, N. 1998. Health and safety—addressing barriers to improved safety performance. *Construction Manager* November: 14-15.

Akilade, A. 2000. What now? *Building* 2 June: 18-20.

Ansoff, H. I . 1979. *Strategic management.* London: Macmillan.

_____. 1984. *Implanting strategic management.* Englewood Cliffs, NJ: Prentice Hall.

Argyris, C. 1984. Double loop learning in organizations. In *Organizational Psychology—Readings on Human Behavior and Communications.* Edited by D.A. Kolb, I.M. Rubin, and McIntyre, J. M. Englewood Cliffs, NJ: Prentice Hall: 45-58.

_____. 1990. *Overcoming organizational defenses.* London: Allyn and Bacon.

Aspery, J. 1993. The media: friend or foe? *Administrator* 2(2): 17-19.

Barlow, J. 1999. Laing: Rooted in the past planning for the future. *Construction Manager* April: 22-24.

Barnes, M. 1991. Risk sharing in contracts. In *Civil Engineering Project Procedure in the EC.* Proceedings of the conference organized by the Institution of Civil Engineers, Heathrow, London, January 24-25.

Barrett, P.S. (Ed.) 1995. Facilities management—towards best practice. London: Blackwell Science.

Bavelas, A. 1950. Communication patterns in task orientated groups. *Acoustical Society of America Journal* 22: 727-30.

Bax, E.H.; B.J. Steijn; and M.C. DeWitte. 1998. Risk management at the shop floor: The perception of formal rules in high-risk work situations. *Journal of Contingencies and Crisis Management* 6(4): 177-88.

Baxendale, A.T. 1991. Management information systems: The use of work breakdown structure in the integration of time and cost. In *Management, Quality, and Economics in Building.* Edited by A. Bezelga and P. Brandon. Transactions of

CIB Symposium on management, quality and economics in housing and other building sectors, Lisbon, September 30 – October 4: 4-23

Bea, R.G. 1994. *The role of human error in design, construction, and reliability of marine structures*. Ship Structure Committee SSC-378, U.S. Coastguard, Washington, DC.

Benini, A.A. and Benini, J.B. 1996. Ebola virus: from medical emergency to complex disaster. *Journal of Contingencies and Crisis Management* 4(1): 10-20.

Bennis, W. 1996. *Facing the challenge: Peter Drucker in conversation with Warren Bennis*. BBC Enterprises, London.

Benson, J. A. 1988. Crisis revisited: an analysis of strategies used by Tylenol in the second tampering episode. *Central States Speech Journal* 39(1): 49-66.

Best, R. L. 1977. *Reconstruction of a tragedy: The Beverly Hills Supper Club Fire*. Southgate, KY: National Fire Protection Association.

Bignell, V. 1977. The West Gate bridge collapse. In *Catastrophic Failures*, edited by V. Bignell, G. Peters, and C. Pym. Milton Keines, UK: The Open University Press, 127-66.

Blockley, D. I. 1996. Hazard engineering, In *Accident and Design: Contemporary Debates in Risk Management*. Edited by C. Hood and D.K. C. Jones. UCL Press, London, 31-38.

Bobo, C. 1997. Hitachi faces crisis with textbook response. *Public Relations Quarterly* 42(2): 18-21.

Booth, S. A. 1993. *Crisis management strategy—competition and change in modern enterprises*. London: Routledge.

Bovens, M. 1996. The integrity of the managerial state. *Journal of Contingencies and Crisis Management* 4(3): 125-33.

British Property Federation. 1983. *Manual of the BPF system—The British Property Federation System for building design and construction*. London: The British Property Federation Ltd.

Brecher, M. 1977. Toward a theory of international crisis behaviour—A preliminary report. *International Studies Quarterly* 21(1): 39-74.

Building. 1997. Image building. *Building* September 19: 18-23.

Bullock, C. 1999. *Killology*. A discussion on Australian Broadcasting Corporation. Sunday, June 2. http://www.abc.net.au/rn/talks/bbing/stories/s23921.htm.

Burns, T. and G. Stalker. 1961. *The management of innovation*. London: Tavistock.

Butterfield, H. 1975. Introduction. In *The Sleepwalkers: A history of man's changing vision of the universe,* edited by A. Koestler. Harmondsworth, UK: Penguin Books.

Camp, R. C. 1989. *Benchmarking: The search for industry best practices that lead to superior performance*. Milwaukee, WI: ASQC Quality Press.

Carmichael, D. 1999. Gurus of faddish management. In *Proceedings of the international conference on construction process re-engineering,* edited by K. Karim, et al. Construction Process Re-engineering, University of New South Wales, Sydney, Australia, July 12-13: 365-74.

Carper, K. L. 1989. What is forensic engineering? In *Forensic Engineering*, edited by K.L. Carper. New York: Elsevier Science Publishing, 1-12.

Cavill, N. 1999. Purging the industry of racism. *Building* 14 May: 20-22.

Cherns, A. B. and D. T. Bryant. 1984. Studying the client's role in construction management. *Construction Management and Economics* 1(2): 177-84.

Cisin, I. H. and W. B. Clark. 1962. The methodological challenge of disaster research. In *Man and society in disaster*, edited by G.W. Baker G W and D.W. Chapman. New York: Basic Books Inc, 23-49.

Clennell, A. 1999. Storm bill biggest in Australia. *The Sydney Herald*, May 8, p. 5.

Comfort, L. K. 1993. Integrating information technology in to international crisis management and policy. *Journal of Contingencies and Crisis Management* 1(1): 15-27.

Commonwealth of Australia. 1999. *Building for growth—An analysis of the Australian building and construction industries*. Canberra, Australia: Australian Government Printing Service.

Construction Management and Economics.1997. Special edition on law and dispute resolution in construction. *Construction Management and Economics* 15(6): 501-75.

Cooper, D. F. and C. B. Chapman. 1987. *Risk analysis for large projects—Models, methods, and cases*. New York: John Wiley and Sons.

Craig, R. 1996. Manslaughter as a result of a workplace fatality. In *ARCOM 96, Proceedings of 12th Annual ARCOM Conference*, September 11-13. Edited by A.Thorpe. Sheffield Hallam University, UK, Volume 1: 1-11.

Daft, R. L. 1983. *Organizational theory and design*. St. Paul, MN: West Publishing Company.

Davis, S. M. 1987. Future perfect New York: Addison Wesley.

Davis, D. T. 1995. Harzardous materials contingency planning. In *Proceedings of 2nd International Conference on Loss Prevention and Safety*, October 16-18. Edited by F. Bushehri. Bahrain Society of Engineers, Bahrain, pp. 503-13.

DeMichiei, J., J. Langton, K. Bullock, and T. Wiles. 1982. Factors associated with disabling injuries in underground coalmines. *Mine Safety and Health Administration* June: 72.

Department of Energy. 1990. The public inquiry into the Piper Alpha disaster. The Honourable Lord Cullen, Volume 1 and Volume 2. London: HMSO.
Department of Public Works. 1999. *Creating an enabling environment for reconstruction, growth and development in the construction industry*. Pretoria, South Africa: Department of Public Works Government Printer.

Dixon, N. F. 1988. *On the psychology of military incompetence*. San Francisco: Cape.

Egelhoff, W. G. and F. Senn. 1992. An information processing model of crisis management. *Management Communication* 5(4): 443-84.

Eldukair, Z. A. and B. M. Ayyub. 1991. Analysis of recent U.S. structural and construction failures. *Journal of Performance of Constructed Facilities* 5(1), 57-73.

Engineering and Construction Contract. 1995. *The engineering and construction contract*. London: Thomas Telford.

Feature. 1999. Lagging funds delay Las Vegas Flood Control Project. *Civil Engineering*, April: 18-19.

Fennelle, C. 1996. The response to TWA Flight 800: Lessons learned. *Risk Management* 43(11): 58-69.

Fink, S. L.; J. Beak; and K. Taddeo. 1971. Organizational crisis and change. *Journal of Applied Behavioral Science* 7(1): 15-37.

Forman, D. 1993. Emergency plans and the mini-crisis. *Business Review Weekly* March 12: 72-73.

Freeman, L. C. 1979. Centrality in social networks conceptual clarification. *Social Networks* 1: 215-39.

Gablentz, O. H. 1972. Responsibility. In *The encyclopaedia of the social sciences, volumes 13 and 14*. Edited by D. L. Sills. New York: The Macmillan and Company and Free Press, 496-500.

Gambatese, J. A. 2000 Safety in a designer's hands, *Civil Engineering*, June: 5659.

Gay, A. 1998. Why trophy winners lose. *Building* October 30: 29.

George, A. L. 1991. Strategies for crisis management. In *Avoiding War—Problems of crisis management*. Edited by A.L. George. San Francisco: Westview Press.

Glackin, M. and G. Barrie. 1998. Jubilee Line strike set to end after last-ditch talks. *Building*. November 27: 12.

Glackin, M. 2000. Wembley row threatens to put back stadium a year. *Building*, 14 April: 10.

Goldberg, S. D. and B. B. Harzog. 1996. Oil spill: Management crisis or crisis management? *Journal of Contingencies and Crisis Management* 4(1): 1-10.

Gonzalez-Herrero, A. and C. B Pratt. 1995. How to manage a crisis before—or whenever—it hits. *Public Relations Quarterly* 40(1): 25-29.

Gray, C.; W. P. Hughes; and J. Bennett. 1994. *The successful management of design—A handbook of building design management*. Reading, U.K.: University of Reading, Center for Strategic Studies.

Green, S. 1998. The technocratic totalitarianism of construction process improvement: a critical perspective. *Engineering, Construction, and Architectural Management* 5(4): 376-86.

Hall, G; J. Rosenthal; and J. Wade. 1993. How to make reengineering really work *Harvard Business Review* November-December: 119-31.

Haroon, S. 1999. *Engineering disaster, the Ford Pinto case: A study in applied ethics, business, and technology*. http://www.uoguelph.ca/ca/~sharoon/a1/A1disaste.htm.

Hatush, Z. and M. Skitmore. 1997. Evaluating contractor prequalification data: selection criteria and project success factors. *Construction Management and Economics* 15(3): 129-47.

Heller, R. 1993. TQM – Not a panacea but a pilgrimage. *Management Today* January: 37-40.

Hemsley, A. 1998. Nightmares of recession. *Building* November 20: 33.

Hendry, R. 1989. Industrial accidents. In *Forensic Engineering*. Edited by K. L. Carper. New York: Elsevier Science Publishing, 56-80.

Health and Safety Commission. 1994. Health and safety statistics: statistical supplement to 1993/94 Annual Report. Sudbury, U.K.: HSE Books.

Hermann, C. F. 1963. Some consequences of crisis which limit the viability of organizations. *Administrative Science Quarterly* 8(25): 61-82.

Hilmer, F. G. and L. Donaldson. 1996. *Management redeemed—Debunking the fads that undermine corporate performance.* London: The Free Press.

Hindle, R. D. and M. H. Muller. 1996. The role of education as an agent of change: A two-fold effect. *Journal of Construction Procurement* 3(1): 56-66.

Horlick-Jones, T. 1996. The problem of blame. In *Accident and design: Contemporary debates in risk management*. Edited by C. Hood and D.K.C. Jones. UCL Press, London, 61-70.

Hornstein, H. A. 1986. *Managerial courage.* New York: John Wiley and Sons.

Irvine, R. B. 1997. *What's a crisis anyway? Surviving a business crisis.* Glanbridge Publishing Ltd, Urbana, Illinois.

Janis, I. L. 1988. Groupthink. In *Behavior in organizations—An experiential approach*. 4th edition. Edited by J. B. Lau and A. B. Shani., Irwin Homewood, Urbana, Illinois, 162 - 69.

Jarman, A. and A. Kouzmin. 1990. Decision pathways from crisis—a contingency theory simulation heuristic for the challenger space disaster (1983-1988). In *Contemporary Crisis –Law, Crime and Social Policy*. Edited by A. Block. Netherlands: Kluwer Academic Press, 399-433.

Johns, J. N. 1999. Letter to editor. *Civil Engineering* May 6, 37.

Kanter, E. 1983. *The change masters*. London,: Allen and Unwin.

Kelly, J.; S. MacPherson; and S. Male. 1992. *The briefing process: A review and critique*. Paper Number 12, The Royal Institution of Chartered Surveyors.

Kelman, H. C. 1997. Social-psychological dimensions of international conflict, In

Khosrowshahi, F. 1996. A neural network model for bankruptcy prediction of contracting organizations. In *ARCOM 96, Proceedings of 12th Annual ARCOM Conference*, September 11-13. Edited by A. Thorpe. Sheffield Hallam University, UK, p. 200-09.

Knight, K. E. and R. R. McDaniel, Jr. 1979. *Organizations—An information systems perspective* London: Wadsworth.
Knutt, E. 1998. Countdown to the millennium. *Building* June 19.

_____. 1998. The troubleshooter. *Building* June 19.

_____. 2000. Are we safer now? *Building* April 7.

Koestler, A. 1975. *The Sleepwalkers: A history of man's changing vision of the universe*. Harmondsworth, UK: Penguin Books.

Kumaraswamy, M. M. 1996. Construction dispute minimisation. In *The Organization and Management of Constructio*. Edited by D.A. Langford, A. Retik. London: E and F N Spon, 447-57.

Kutner, M. 1996. Coping with crisis. *Occupational Health and Safety* 65(2): 22-24.

Latham, M. 1994. *Construction the team*. Final Report of the Government/Industry Review of Procurement and Contractual Arrangements in the UK Construction Industry. London: HMSO.

Leavitt, H. J. 1951. Some effects of certain communication patterns on group performance. *Journal of Abnormal Social Psychology* 46(1): 38-50.

Leavitt, H. J. and H. Bahrami. 1988. *Managerial psychology—managing behavior in organizations* 5th ed. Chicago: The University of Chicago Press.

Likert, R. 1961. *New patterns of management*. New York: McGraw-Hill.

Loosemore, M. 1999. The problem with business fads. In *Proceedings of the international conference on construction process re-engineering*. Edited by K. Karim, et al. Construction Process Re-engineering. University of New South Wales, Sydney, Australia, July 12-13: 355-63.

Loosemore, M. and K. Hughes. 1998. Emergency systems in construction contracts. *Engineering, Construction, and Architectural Management* 5(2):189-99.

Loosemore, M. and T. Chin Chin. 2000. Occupational stereotypes in the construction industry. *Construction Management and Economics*. 18 (5), 559-567.

Loosemore, M.; B. T. Nguyen; N. Dennis. 1999 An investigation into the merits of encouraging conflict in the construction industry. *Construction Management and Economics*, 18 (4), 447-457.

Marsh, P.; E. Roser; and R. Harre'. 1978. *The rules of disorder*. London: Routledge.

Marshall, J. 1997. The professionals. *Building*. December 5: 22-23.

Mazur, L. 1996. Confidence in crisis. *Marketing* R24-R25, June 6.

Miles, L 1967. *Techniques of value analysis and engineering*. New York: McGraw Hill Book Company.

Mincks, W. R. 1996. Construction waste management: reducing and recycling construction and demolition waste. http://www.arch.wsu.edu/bminks/cons.htm.

Mintzberg, H. 1976. Planning on the left side and managing on the right side. Harvard Business Review 54(2): 49-58.

Mintzberg, H. 1979. *The structuring of organizations*. Englewood Cliffs, NJ: Prentice-Hall.

Mitroff, I. and C. Pearson. 1993. *Crisis management: A diagnostic guide for improving your organization's crisis preparedness*. San Francisco: Jossey-Bass Publishers.

Moodley, K and C.N. Preece. 1996. Implementing community policies in the construction industry. In *The Organization and Management of Construction*. Edited by D.A. Langford and A. Retik. London: E and F N Spon: 178-86.

Morley, I. E. 1981. Bargaining and negotiation. In *Psychology and Management*. Edited by C. L. Cooper. Macmillan Press Ltd, New York, 95-127.

Morris, P. 1998. Warnings for designers. *Building* April 9: 32-33.

Morris, P and G. H. Hough. 1987. *The anatomy of major projects—A study of the reality of project management*. Chicester, UK: John Wiley and Sons.

Munns A. K. 1996. Measuring mutual confidence in UK construction projects. *ASCE Journal of Management in Engineering* 12 (1): 26- 33.

National Bureau of Standards. May 1982. *Investigation of the Kansas City Hyatt Regency walkways collapse.* Building Science Series 143. Washington, DC: National Bureau of Standards, U S Department of Commerce.

National Economic Development Council. 1983. *Faster building for industry.*HMSO: London.

Nicodemus, J. 1997. Operational crisis management. *Secured Lender* 53(6): 84-90.

Oberlender, G. D. 1993. *Project management for engineering and construction.* New York: McGraw-Hill, Inc.

Oliver, J. 1993. Shocking to the core *Management Today* August: 18-22.

O'Rouke, J. 1998. *Union's plea: racism not all right on site.* Syendy Herald Sun, October 11, 19-20.

Pascale, R. T. 1991. *Managing on the edge.* Harmondsworth, UK: Penguin Books.

Pearson, C. M. and J. A. Clair. 1998. Reframing crisis management. *Academy of Management Review* 23(1): 59-76.

Pearson, C. M.; S. K. Misra; J. A. Clair; and I. Mitroff. 1997. Managing the unthinkable. *Organizational Dynamics* 26(2): 51-64.

Perry, J. G. and R. W. Hayes. 1985. *Risk and its management in construction projects.* Proceedings of Institute of Civil Engineers 1(78): 499-521.

Philips, R. C. 1988 Managing changes before they destroy your business: A non-traditional approach, *Training and Development Journal* 42(9): 66-71.

Pinnell, S. 1999. Resolution solutions. *Civil Engineering* June: 62-63.

Pritzker, P. E. 1989. Fire investigation. In *Forensic Engineering.* Edited by K. L. Carper. New York: Elsevier Science Publishing, 32-53.

Quarantelli, E. L. 1993. Community crises: An exploratory comparison of the characteristics and consequences of disasters and riots. *Journal of Contingencies and Crisis Management* 1(2): 167-79.
Reid, J. 1999. Engineers' top 10 crises. *Civil Engineering* April: 27.

Richardson, W. 1996. Modern management's role in the demise of a sustainable society. *Journal of Contingencies and Crisis Management* 4(1): 20-31.

_____. 1993. Identifying the cultural causes of disaster: An analysis of the Hillsborough Football Stadium Disaster. *Journal of Contingencies and Crisis Management* 1(1): 27-36.

Robertson, I. T. and C. L. Cooper. 1983. *Human behavior in organizations.* London: MacDonald and Evans.

Rochlin, G. I., ed. 1996. New directions in reliable organization research. *Journal of Contingencies and Crisis Management* 4(2).

Rogers, E. M. and D. L. Kincaid. 1981. *Communication networks. Toward a new paradigm for research.* London: The Free Press.

Rogers, J. P. 1991. Crisis bargaining codes and crisis management. In *Avoiding War—Problems of crisis management.* Edited by A. L. George. San Fransisco: Westview Press, 413-42.

Rosenthal, U. and A. Kouzmin. 1993. Globalization: An addenda for contingencies and crisis management—An editorial statement. *Journal of Contingencies and Crisis Management* 1(1): 1-11.

Ryan, K. D. and D. K. Oestreich. 1998. *Driving fear out of the workplace,* 2nd ed. San Francisco: Jossey-Bass Publishers.

Sagan, S. D. 1993. *The limits of safety: Organizations, accidents and nuclear weapons.* Princeton, NJ: Princeton University Press.

_____. 1991. Rules of engagement. In *Avoiding war— Problems of crisis management.* Edited by A. L. George. San Francisco: Westview Press, 443-70.

Scott, J. 1991. *Social network analysis—A handbook.* London:Sage Publications, Ltd.

Sfiligoj, E. 1997. In the midst of a crisis. *Beverage World* 116(1): 92-107.

Shaw, M. E. 1954. Group structure and the behavior of individuals in small groups. *Journal of Psychology* 38: 139-49.

Sheaffer Z.; B. Richardson and Z. Rosenblatt. 1998. Early warning signals management: A lesson from the bearings crisis. *Journal of Contingencies and Crisis Management* 6(1): 1-23.

Sheldrake, J. 1996. *Management theory—From Taylorism to Japanization.* London: Thompson Business Press.

Shrivastava, P. 1992. *Bhopal: Anatomy of a crisis*. London: Paul Chapman Publishing.

Sipika, C. and D. Smith. 1993. Back from the brink—post crisis management. *Long Range Planning* 1(1): 28-38.

Sinclair, A. and F. Haines. 1993. Deaths in the workplace and the dynamics of response. *Journal of Contingencies and Crisis Management* 1(3): 125-38.

Smith, C. 1996. High risk business. *Airline Business* November: 88-89.

Smith, N. J. 1999. *Managing risk in construction projects*. Oxford, UK: Blackwell Science.

Snyder, G. H. 1972. Crisis bargaining. In *International crisis—Insights from behavioral research*. Edited by C.F. Hermann. London: The Free Press, Collier-Macmillan Ltd, 217-56.

Spring, M. 1998. Gentle Giant *Building* October 31: 41-45.

Stacey, R. 1992. *Managing chaos*. London: Kogan Page.

Stewart, V. 1983. *Change—The challenge of management*. New York: McGraw-Hill.

Teo, M. M. M. 1998. *An investigation of the crisis-preparedness of Australian Construction Companies*. Unpublished BSc Thesis, University of New South Wales, Sydney, Australia.

t'Hart, P. 1993. Symbols, rituals and power: The lost dimensions of crisis management. *Journal of Contingencies and Crisis management* 1(1): 36-51.

Theodore, J. 1975. *The Empire State Building*. New York: Harper and Row.

Tichy, N. M.; M.L. Tushman; and C. Fombrun. 1979. Social network analysis for organizations. *Academy of Management Review* 4(4): 507-19.

Toffler, A. 1970. *Future Shock— A study of mass bewilderment in the face of accelerating change*. London: The Bodley Head.

Tsoukas, H. 1995. *New thinking in organizational behavior*. Oxford, UK: Butterworth-Heinnemann.

Tsurumi, R. 1982. American origins of Japanese productivity: The Hawthorne Experiment rejected. *Pacific Basin Quarterly* 7(1): 14-15.

Turner, B. A. and N F. Pigeon. 1997. *Man-made Disasters*, 2nd ed. Oxford, UK: Butterworth Heinemann.

Uff, J. 1995. Contract documents and the division of risk. In *Risk Management and Procurement in Construction*. Edited by J. Uff and A. M. Odams. London: Centre for Construction Law and Management, Kings College, 49-69

Wagenaar, W. A. 1996. Profiling crisis management. *Journal of Contingencies and Crisis Management* 4(3): 169-74.

Watson, T. J. 1994. *In search of management*. London: Routledge.

Weyer, M. V. 1994. When the wheels come off. *Management Today* July: 30-37.

Wildavsky, A. 1988. *Searching for safety*. New Brunswick: Transaction.

Wills, J. A. 1996. Enhancing disaster planning and management: The 1990 Nyngan flood in Australia. *Journal of Contingencies and Crisis Management* 4(1): 32-39.

Wire Headlines. 1999. *Long prison terms given in South Korean Mall Collapse*. http://sddt.com/files/librarywire/96wireheadlines/08-96/DN96-08-23/DN96-08-23-1html.

Zartman, I. W. and J. L. Rasmussen. 1997. *Peace making in international conflict: Methods and techniques*. Washington, DC: United States Institute of Peace.

INDEX